Viral Crossroads: The Interface Between Wildlife and Domestic Animal Health

Viral Crossroads: The Interface Between Wildlife and Domestic Animal Health

Guest Editors

Vesna Milicevic
Teufik Goletić

MDPI

Basel • Beijing • Wuhan • Barcelona • Belgrade • Novi Sad • Cluj • Manchester

Guest Editors
Vesna Milicevic
Virology Department
Institute of Veterinary
Medicine of Serbia
Belgrade
Serbia

Teufik Goletić
Department of Pathobiology
and Epidemiology
University of
Sarajevo—Veterinary Faculty
Sarajevo
Bosnia and Herzegovina

Editorial Office
MDPI AG
Grosspeteranlage 5
4052 Basel, Switzerland

This is a reprint of the Special Issue, published open access by the journal *Veterinary Sciences* (ISSN 2306-7381), freely accessible at: https://www.mdpi.com/journal/vetsci/special_issues/087V9K27RC.

For citation purposes, cite each article independently as indicated on the article page online and as indicated below:

Lastname, A.A.; Lastname, B.B. Article Title. *Journal Name* **Year**, *Volume Number*, Page Range.

ISBN 978-3-7258-6109-5 (Hbk)
ISBN 978-3-7258-6110-1 (PDF)
https://doi.org/10.3390/books978-3-7258-6110-1

Contents

veterinary sciences

MDPI

Article

Endemic Circulation of Cluster 19 African Swine Fever Virus in Serbia and Bosnia and Herzegovina

Dimitrije Glišić [1,*,†], Šejla Goletić Imamović [2,†], Sofija Šolaja [1], Ilma Terzić [2], Ajla Hodžić Borić [2], Teufik Goletić [2] and Vesna Milicevic [1]

1 Department of Virology, Institute of Veterinary Medicine of Serbia, Janisa Janulisa 14, 11000 Belgrade, Serbia; sofija.solaja@nivs.rs (S.Š.); vesna.milicevic@nivs.rs (V.M.)
2 Veterinary Faculty, University of Sarajevo, Zmaja od Bosne 90, 71000 Sarajevo, Bosnia and Herzegovina; sejla.goletic@vfs.unsa.ba (Š.G.I.); ilma.terzic@vfs.unsa.ba (I.T.); ajla.hodzic.boric@vfs.unsa.ba (A.H.B.); teufik.goletic@vfs.unsa.ba (T.G.)
* Correspondence: dimitrije.glisic@nivs.rs
† These authors contributed equally to this work.

Simple Summary

African swine fever is a deadly viral disease of pigs and wild boar that causes major losses for farmers and threatens food security. The disease does not affect people, but its rapid spread and high fatality in pigs make it one of the most serious challenges for animal health in Europe. Since 2019, the disease has been present in Serbia, and in 2023, it was first reported in Bosnia and Herzegovina. In this study, we examined virus samples collected from pigs and wild boar during outbreaks between 2023 and 2025 to better understand how the virus is spreading in the region. By looking at several important parts of the virus genome, we found that all the samples belonged to the same group, known as cluster 19. This shows that the same type of virus has been circulating for several years without major changes. The results suggest that the disease is being maintained locally, mainly through contact between wild boar and pigs kept on small farms with little or no protection. The discovery of the same virus type in Bosnia and Herzegovina highlights that the disease crosses borders, making regional cooperation and continued monitoring essential for controlling its spread.

Abstract

African swine fever (ASF) is a highly fatal viral disease of domestic pigs and wild boar that continues to threaten pig production across Europe. Genotype II African swine fever virus (ASFV) has been present in Serbia since 2019 and was first confirmed in Bosnia and Herzegovina in 2023, yet recent genetic data from the region have been lacking. This study aimed to update the genetic characterization of ASFV strains circulating in Serbia between 2023 and 2025 and to provide the first sequence data from Bosnia and Herzegovina. A total of 110 isolates were analyzed by partial sequencing of seven genomic regions recommended by the European Union Reference Laboratory. Good-quality sequences were obtained for at least two loci per isolate. All isolates belonged to genotype II and were classified as CVR variant I, IGR-II, O174L-I, MGF I, K145R-I, and ECO2-II, corresponding to cluster 19. No novel genetic changes were identified in the sequenced fragments. These findings indicate the stable, endemic circulation of cluster 19 in both domestic pigs and wild boar, maintained through ecological and human-mediated transmission at the wildlife–livestock interface. The detection of cluster 19 in Bosnia and Herzegovina underscores transboundary spread and highlights the need for continued molecular surveillance and regional cooperation.

Keywords: African swine fever; genetic analysis; genotyping; domestic pig; wild boar; Serbia; Bosnia and Herzegovina

1. Introduction

African swine fever (ASF) is a highly fatal viral haemorrhagic disease affecting members of the *Sus scrofa* species. The causative agent, African swine fever virus (ASFV), is the sole member of the *Asfarviridae* family and the *Asfivirus* genus [1]. ASFV has a linear double-stranded DNA genome ranging from 170 to 190 kbp in length, depending on the genotype, subgenotype, and the number of repeat sequences. The genome encodes 150–160 open reading frames, capable of producing over 150 structural and non-structural viral proteins [2]. To date, 24 viral genotypes have been identified, all originating in Africa [2]. However, a recent study showed that genotype XVIII, previously represented by a single Namibian isolate, was a composite of genotypes I and VIII, and its retirement has been recommended to avoid future confusion [3]. Only two genotypes have spread beyond the African continent: genotype I, which caused outbreaks in Europe and the Americas during the 20th century, and genotype II, responsible for the ongoing global pandemic [4]. Genotype I was eradicated from mainland Europe by 1995, but persisted on the island of Sardinia, where it is now considered close to eradication, marking a significant milestone in European ASF control efforts [4].

ASF was first confirmed in Serbia in 2019 in domestic pigs and in 2020 in wild boar [5]. Since then, case numbers have steadily increased in both populations. The highest number of outbreaks was recorded in 2023, with 992 in domestic pigs and 213 in wild boar [6]. In 2024, the numbers dropped to 310 outbreaks in domestic pigs and 101 in wild boar [7]. Bosnia and Herzegovina reported its first case in 2023. That same year, it recorded 1511 outbreaks in domestic pigs and 29 cases in wild boar. In 2024, numbers fell to 33 outbreaks in domestic pigs and 38 in wild boar [6,7]. In 2025, the trend continues downward. Serbia has reported 20 outbreaks in domestic pigs and 28 in wild boar. Bosnia and Herzegovina has reported 8 outbreaks in domestic pigs and 15 in wild boar [8]. In Serbia and Bosnia and Herzegovina, pig farming is largely based on extensive systems [9]. Over 40% of registered farms are smallholdings with low or no biosecurity [10]. Most are family-owned and keep a sow with piglets for personal use. In some regions, pigs are seasonally pastured, increasing contact with wild boar. In other areas, domestic pig–wild boar hybrids are kept and valued as a delicacy, allowing for bidirectional contact in the domestic pig-wild boar interface [11,12]. Since the introduction of ASF, the domestic pig population has decreased significantly. Between 2020 and 2024, the pig population in Serbia declined by approximately 25%, while the European Union (EU-27) experienced a reduction of about 10% over the same period, driven by ASF outbreaks, high production costs, and reduced market demand [13–15].

Genotype II ASFV strains in Europe have been divided into over 24 genetic clusters based on multi-gene sequencing, as recommended by the European Reference Laboratory (EURL) [4,16]. Genotyping of ASFV is primarily based on partial sequencing of the B646L gene, which distinguishes 23 known genotypes [3,17]. For deeper classification within genotype II, additional markers are used. The CVR region of the B602L gene, based on the amino acid tandem repeat sequences (TRS), defines 31 subgroups in Africa and two major variants in Europe, with further diversity identified through single-nucleotide polymorphisms (SNPs). Intergenic regions (IGR) also contribute to sub-grouping, particularly the ECO1 region (IGR between I73R and I329L), where four variants (IGR I–IV) are defined by the TRS "TATATAGGAA", with IGR II being the most common in Europe. The O174L gene allows the distinction of two variants based on the TRS "CAGTAGTGATTTTT", while TRS in the MGF 505 9R/10R IGR ("AGTAGTTCAGTTAAGAT") define eight variants, with variant I dominating in European strains. Further differentiation is possible through SNP analysis in the K145R and I215L (ECO2) genes, which define four and two variants, respectively [4]. Based on the multi-gene approach, cluster 19—defined by the markers CVR-1, IGR-II, O174L-I, K145R-I, MGF-I, and ECO2-II—is the dominant variant in Bulgaria, Serbia, Greece, and North Macedonia [4]. In Serbia, additional variations in the CVR and IGR regions have been identified, suggesting the presence of other clusters [16]. In 2023, outbreaks were also reported in Croatia and North Macedonia [6]. Previous spatio-temporal analyses of ASF-infected wild boar showed clusters extending beyond national borders, suggesting regional circulation of related strains [10].

Although ASF has been circulating in Serbia since 2019 and was first detected in Bosnia and Herzegovina in 2023, almost no molecular data have been generated for either country in the past three years. As a result, it is unclear whether the current outbreaks are caused by the continued circulation of previously introduced strains or by new incursions from neighbouring regions. This uncertainty limits veterinary services' ability to trace transmission routes, evaluate control measures, and detect emerging viral variants. The need for updated genetic data is particularly relevant in the Balkans, where small-scale pig farming, low biosecurity, and frequent cross-border movement of pigs and pork products create favourable conditions for virus persistence. Molecular epidemiology helps determine whether ASFV outbreaks result from new introductions or continued local circulation. Although genotype II ASFV shows minimal genetic variation, sequencing of several genomic regions still allows classification into more than 24 clusters across Europe [4]. The virus evolves slowly, and because cluster 19 predominates in Southeast Europe, tracing transmission routes can be difficult using routine markers. Even so, regions such as the CVR in B602L, the ECO1 and ECO2 intergenic regions, and O174L remain useful for detecting minor differences and distinguishing between persistence and new incursions [4,18]. Beyond the structural gaps in surveillance and biosecurity, everyday human activities play a major role in keeping ASF circulating in the Balkans [9]. The disease often spreads quietly through informal trade of pigs and pork products, home slaughtering, and the delayed reporting of sick or dead animals. This situation is not unique to Serbia and Bosnia and Herzegovina; in Romania and Bulgaria, the highest number of outbreaks has also been recorded in backyard farms with poor biosecurity [19]. In many rural areas, local cross-border movements of pigs, carcasses or equipment take place with minimal control, allowing the virus to move freely between communities and across national borders. Together, these social and economic realities create conditions that allow the virus to persist, even when official outbreak numbers begin to drop. Earlier studies from Serbia reported that genotype II strains belonged mainly to cluster 19, with occasional variation in the CVR and intergenic regions, suggesting microevolution or multiple introductions. This study aims to address the current gap by sequencing ASFV samples from domestic pigs and wild boar collected during outbreaks in Serbia and Bosnia and Herzegovina between 2023 and 2025.

2. Materials and Methods

A total of 69 samples from Serbia and 40 samples from Bosnia and Herzegovina were subjected to total nucleic acid extraction, RT-qPCR and subsequent sequencing. Samples from Bosnia and Herzegovina included 34 spleen samples (19 from wild boar and 15 from domestic pigs), three whole blood samples and three serum samples (domestic pigs), while samples from Serbia included 67 blood swabs (domestic pigs) and two spleen samples (wild boar). All samples were collected as part of official ASF surveillance and outbreak control activities, in compliance with national veterinary legislation in Serbia and Bosnia and Herzegovina. As the material was obtained within these compulsory control measures and not for experimental purposes, additional ethical approval was not required [20,21]. Spleen samples, cut into approximately 0.1 g pieces, were homogenized in 1000 μL PBS using a metal bead, while each swab was placed in 1 mL of PBS and homogenized with a TissueLyser II (Qiagen, Hilden, Germany) for 5 min at 30 Hz. The samples were then briefly centrifuged, and the supernatant was used for further extraction steps using the DNeasy Blood & Tissue (Qiagen, Hilden, Germany) according to the manufacturer's instructions. Whole blood samples and serum samples were extracted using the Purification of Total DNA from Animal Blood or Cells protocol of the DNeasy Blood & Tissue Kit (Qiagen, Hilden, Germany) according to the manufacturer's instructions. Any remaining tissue samples were stored at −80 °C. Nucleic acid extraction from samples collected in Serbia was performed using the IndiSpin Pathogen Kit (Indical Bioscience GmbH, Leipzig, Germany), following the manufacturer's protocol. To confirm the presence of the viral RNA, an RT-qPCR protocol by King et al. [22] was used as described by Protocols and Guidelines for Laboratory Diagnosis of ASFV by WOAH [23].

The samples were chosen for sequencing based on their Ct value (≤27) to ensure sufficient template quantity, origin (wild/domestic pig) and the time of sampling to ensure that representative data for outbreaks 2023–2025 would be obtained. As per recommendation by the EURL, targeted sequencing was performed using primers targeting seven regions of ASFV genome: 487 bp region of the B646L locus [17], 491 bp region of the CVR of B602L locus [24] 356 bp TRS present in the IGR between I73R and I329L (ECO1) [25], 637 bp region of O174 locus [26], 282 bp region of K145R locus [26], 551 bp TRS located in the IGR between the MGF505 9R and 10R loci (MGF) [4] and 604 bp fragment containing segments of I329L locus, I215L locus and IGR between them (ECO2) [4].

PCR amplification conditions differed slightly between genomic regions. For the B646L locus, the reaction began with an initial denaturation at 95 °C for 20 s, followed by 40 cycles of 95 °C for 12 s (denaturation), 50 °C for 20 s (annealing), and 70 °C for 25 s (extension). A final extension step was performed at 72 °C for 10 min. For the ECO1 locus, amplification was carried out using 40 cycles of 95 °C for 30 s, 60 °C for 30 s, and 72 °C for 1 min, with a final extension at 72 °C for 10 min. For the B602L, MGF, and ECO2 regions, the cycling parameters were identical, except that the annealing temperatures were 55 °C for B602L and MGF, and 56 °C for ECO2. For the K145R and O174L loci, PCR conditions followed previously published protocols and consisted of an initial denaturation at 95 °C for 10 min, followed by 40 cycles of denaturation at 95 °C for 15 s, annealing at 50 °C for 30 s, and extension at 72 °C for 30 s. A final elongation was performed at 72 °C for 3 min [17,24–26]. The quantification of DNA was performed using a Qubit dsDNA HS (High Sensitivity) Assay Kit on a Qubit 4 device (Thermo Fisher Scientific, Waltham, MA, USA), following the manufacturers' instructions. For samples from Bosnia and Herzegovina sequencing of the obtained amplicons was performed using Native Barcoding Kit 24 V14 (SQK-NBD114.24, Oxford Nanopore Technologies, Oxford, UK) according to the manufacturer's instructions, on an R10.4.1 flow cell using the MinION Mk1C instrument (Oxford Nanopore Technologies, Oxford, UK). Sequences were basecalled in real time through MinKNOW v.20.10.3 using the high accuracy setting and a quality score set to 9 during the 12 h runtime in order to obtain high-quality coverage of the sequences. Seqkit v.2.8.2 (accessed on the 22 September 2025 https://bioinf.shenwei.me/seqkit/) was used to additionally check sequence quality, after which the reads were mapped to their respective loci on the Georgia 2007/1 reference genome (GenBank No.: FR682468.2) with Minimap2 v2.29-r1283 (accessed on the 22 September 2025 https://github.com/lh3/minimap2/releases). The sequences were then checked and indexed using Samtools 1.21 (accessed on the 22 September 2025 https://github.com/samtools/samtools/releases/), mappings were visually analyzed using UGENE v52.0 [27] and consensus sequences were created with iVar v1.4.3 (accessed on the 22 September 2025 https://github.com/andersen-lab/ivar). PCR products obtained from Serbian samples were purified using the GeneJET PCR Purification Kit (Thermo Fisher Scientific, Waltham, MA, USA) and subsequently sequenced by Sanger sequencing at LGC, Biosearch Technologies (Berlin, Germany). The obtained sequences were run against the NCBI database with the BLAST tool 2.17.0 (accessed on the 22 September 2025 https://blast.ncbi.nlm.nih.gov/Blast.cgi) to confirm genetic similarities with other ASFV sequences and determine their genotype in accordance with EURL standards. All sequences were deposited in GenBank (Table S1).

Finally, the epidemiological data of sequenced samples were used to generate the map of the localities with ASF cases in domestic pigs and wild boar.

3. Results

Sequencing Success and Genetic Classification of ASFV Isolates

From the 69 ASFV isolates originating from Serbia and selected for sequencing, all were successfully amplified on gel PCR; however, good-quality nucleotide sequences were obtained for 67 isolates at the B646L locus (97.1%), 59 at the B602L locus (85.5%), 61 in the ECO I region (88.4%), 68 in the ECO II region (98.6%), 54 at the O174L locus (78.3%), 65 at the K145R locus (94.2%), and 64 in the MGF region (92.8%). Each isolate yielded at least two genomic regions with reliable sequence data, allowing their inclusion in downstream comparative analyses (Table S1). All 40 ASFV isolates from Bosnia and Herzegovina yielded good-quality sequences of all seven targeted regions, and as such, all were used for downstream analysis. Alignment of the B646L gene confirmed that all isolates belonged to genotype II. Further classification followed the EURL recommendations. Analysis of the CVR amino acid TRS (BNDBNDBNAA) revealed 100% identity with the reference strain Georgia 2007/1 (FR682468.1), classifying all isolates as CVR variant I. Examination of the TRS "TATATAGGAA" in the ECO1 region classified all isolates as IGR-II, while analysis of the O174L TRS placed them within O174L-I. Similarly, the TRS within the MGF 505 9R/10R IGR ("AGTAGTTCAGTTAAGAT") classified the isolates as MGF-I. SNP analysis in the K145R and I215L genes further grouped the isolates as K145R-I and ECO2-II, respectively. According to the EURL guidelines, these combined markers place all strains sequenced between 2023 and 2025 into cluster 19.

The spatial distribution of ASF-positive cases in domestic pigs and wild boar in Serbia and Bosnia and Herzegovina over the 2023–2025 period is shown in Figure 1.

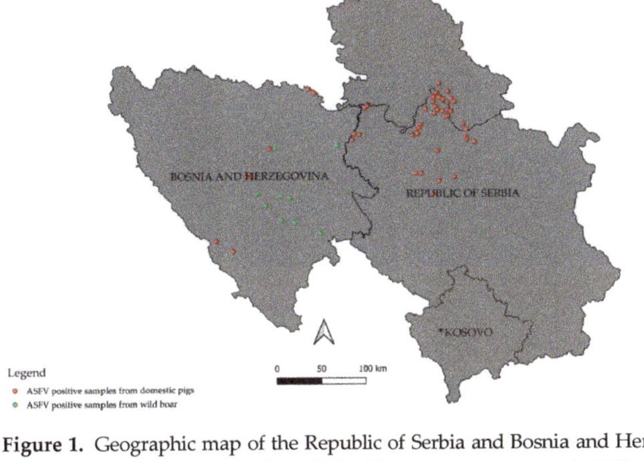

Figure 1. Geographic map of the Republic of Serbia and Bosnia and Herzegovina. Red dots indicate domestic pig samples positive for ASFV, while green dots represent wild boar samples positive for ASFV.

* This designation is without prejudice to positions on status and is in accordance with United Nations Security Council Resolution 1244 (1999) and the advisory opinion of the International Court of Justice on the Kosovo Declaration of Independence.

4. Discussion

This study provides the first updated genetic characterization of ASFV strains circulating in Serbia during 2023–2025 and, importantly, presents the first sequence data from Bosnia and Herzegovina.

The results confirm the continued dominance of cluster 19 in the Balkan region, consistent with earlier reports, and underline the need for sustained molecular surveillance to monitor ASFV dynamics [4]. Elsewhere in Europe, greater diversity has been reported. Romania, for example, shows the highest level of genetic variability, with Gallardo et al. (2023) identifying six distinct ASFV clusters, some apparently unique to that country [4]. At a broader European scale, the highest variability has been associated with the O174L-II variant, which is found in Romania, Poland, and Germany, allowing for fine regional tracking of different strains [4,18]. This variant is associated with changes in the PolX DNA polymerase, which may contribute to increased mutation rates, as well as allowing for regional mapping of strains [18]. By contrast, our findings from Serbia and Bosnia and Herzegovina revealed no such variability; all sequenced isolates were classified as cluster 19, suggesting a genetically stable viral population over the study period. Previous investigations in Serbia had demonstrated a more complex scene. Both ECO2-I and ECO2-II variants were detected, with ECO2-I linked to a secondary introduction of the virus and ECO2-II to the initial incursion [28]. These results were interpreted as evidence of at least two independent introductions, possibly from Romania (ECO2-I) and Bulgaria (ECO2-II). Other studies also identified variation in the ECO1 region and in the CVR of the B602L gene, including the first detection of ECO1 IGR-III and several CVR-specific SNP variants [16]. In the present dataset, none of these were observed. The most likely explanation is that such variants either faded out without wider spread or represented restricted circulation events that were only captured incidentally. Similar findings have been described in the Baltic countries, where ASFV has persisted for more than a decade without the emergence of the high number of clusters reported in neighbouring Poland [4].

No novel genetic changes were observed in this study, reinforcing the view that cluster 19 remains the only strain circulating in Serbia and Bosnia and Herzegovina. It should be noted that the statement "no new genetic changes were observed" refers exclusively to the seven genomic regions analyzed. The study did not employ whole-genome sequencing; therefore, mutations outside these loci could not be assessed, and genome completeness or viral infectivity cannot be inferred from the present data. ASFV has now been established in both domestic pigs and wild boar in Serbia for more than five years and can be considered endemic. Although case numbers have declined since 2023, sporadic outbreaks continue to be reported in both countries, indicating that the virus remains present in local populations. Transmission is most likely sustained through interactions at the wild boar–human–domestic pig interface, particularly in backyard farms where biosecurity is minimal or absent. Practices such as unregulated pig movements (e.g., unregulated trade) and the illegal disposal of carcasses further maintain infection pressure in wild boar populations. The persistence of a single stable cluster over several years suggests that once established, ASFV can be maintained without repeated introductions, largely through ecological and human-mediated cycles. Although our molecular findings confirm the continued circulation of a single ASFV cluster, they alone cannot explain how the virus spreads. Considering the close cultural and trade connections between Serbia and Bosnia and Herzegovina, the movement of pigs and pig products across the border likely plays a role in maintaining transmission.

In wild boar populations, ASFV can be maintained independently of domestic pigs, primarily through carcass-mediated transmission. Infected carcasses act as long-term sources of infection, and although their detection and removal are mandated, reporting by hunters and foresters is often inconsistent or delayed, as it may not align with their personal or economic interests [29]. Experimental and field data show that ASFV remains viable in bone marrow and lymphatic tissues for weeks to months, especially in colder weather or shaded forest environments. Field observations show that wild boar frequently investigate and return to carcasses of conspecifics, which increases the likelihood of contact with infected tissue or contaminated surroundings [30]. In stakeholder surveys, hunters and foresters reported that carcass reporting and removal are sometimes delayed or inconsistently performed due to lack of incentives and practical constraints [29]. In combination with practices such as supplementary feeding or aggregation, these factors contribute to sustained circulation of ASFV within wild boar populations, even in the absence of domestic pig involvement [30]. If carcasses are not located and removed, healthy wild boar may become infected while scavenging or through contaminated soil and vegetation [30]. Practices such as supplementary feeding, delayed carcass disposal and uncontrolled movement of wild boar across administrative borders further support this cycle of persistence. These ecological conditions help explain why ASF continues to circulate even when genetic analyses suggest a stable, unchanged viral population. Future studies that combine molecular data with field investigations of farming practices and animal trade would help clarify these dynamics more fully. As several analyzed loci, including ECO1 and the MGF 505 IGR, contain repetitive sequences that hinder reliable alignment, and given the >99.9% sequence identity reported among genotype II ASFV isolates across Europe, phylogenetic analysis was not conducted, as it would not yield meaningful resolution beyond cluster assignment.

The situation in Bosnia and Herzegovina illustrates the transboundary nature of ASFV spread. The first case in domestic pigs was reported in 2023 in the Republic of Srpska, bordering Serbia [6]. Even though the ASF outbreaks in domestic pigs in Bosnia and Herzegovina remain focused around several localities, the affected wild boar are present throughout the central and eastern parts of the country. This likely enables the spread of the virus between localities both in the country and between two countries (Figure 1). Additionally, many cases likely remain unreported, and the number of positive cases is likely higher. The detection of cluster 19 in both countries provides strong evidence of cross-border circulation and emphasizes the importance of regional cooperation. In Serbia and Bosnia and Herzegovina, ASF control is based on the immediate culling of affected backyard herds, strict movement restrictions, and active carcass collection and disposal, together with efforts to improve farm biosecurity. These measures largely explain the absence of pathological findings and prevent any reliable estimation of natural mortality in the present study [20,21]. Without coordinated surveillance and control, the virus will continue to move across national borders, undermining individual countries' efforts. This study is limited by the sequencing of only a subset of outbreaks and by the lower resolution of targeted sequencing compared to whole-genome approaches. The cost-effectiveness of this sequencing approach is questionable, as sequencing seven genome fragments per sample is labour-intensive and often yields largely identical sequences that provide limited regional resolution. Still, mutations in certain loci, such as the O174L segment noted earlier, may offer valuable discriminatory power for more precise regional tracking of circulating strains. Identifying which genome segments are most informative for ASFV, however, remains challenging. Whole-genome sequencing would, in principle, provide the most reliable insight, yet it is both costly and labour-intensive, and in practice, contributes little additional information given that most genotype II ASFV genomes share over 99.9% identity [31]. Additionally, whole genome sequencing would enable tracking the evolution of the entire genome of ASFV, and this may especially be of great importance in the context of persistence of ASFV in the population of wild boar. This approach would also enable the determination of the origin of outbreaks in domestic pigs. Future work should expand outbreak sampling and incorporate whole-genome sequencing to capture finer-scale variation and to improve detection of potential introductions of novel strains. While both Sanger and targeted sequencing lack the resolution of next-generation approaches, they remain useful for cluster assignment and for confirming the absence of new variants during the study period. In our case, even though the same primers were used, Sanger sequencing yielded incomplete coverage across several regions, whereas the targeted MinION approach produced consistent results throughout, which may indicate a greater robustness of nanopore sequencing to template quality and reaction variability. Occasional sequencing failures in the Serbian set likely arose from reduced template quality or DNA degradation during transport, rather than from genomic deletions, as the targeted ASFV loci are essential and highly conserved. Overall, our findings demonstrate the continued dominance of cluster 19 in Serbia and Bosnia and Herzegovina during 2023–2025. Sustained molecular surveillance, combined with cross-border epidemiological coordination, will be essential for tracking ASFV evolution in the region and for informing effective control and eradication strategies.

5. Conclusions

This study provides updated genetic data on African swine fever virus from Serbia and the first sequencing results from Bosnia and Herzegovina. All analyzed isolates belonged to cluster 19, confirming its continued dominance and stable circulation in the region between 2023 and 2025. The absence of new variants suggests limited introductions from outside and points to endemic maintenance through the wild boar–domestic pig interface. The detection of the same cluster in both countries highlights the cross-border nature of the disease and the need for ongoing molecular surveillance and regional cooperation. However, our results also show the limits of the seven-fragment sequencing approach: in a genetically stable setting such as cluster 19, this method does not provide enough resolution to track individual isolates or local transmission routes. In these cases, whole-genome sequencing or markers with greater variability are needed to gain more detailed epidemiological insights. In addition, strengthening data sharing between laboratories in the region, alongside harmonized sampling strategies for both domestic pigs and wild boar, would improve the early detection of any emerging variants. A more systematic inclusion of wild boar carcasses, particularly in areas with declining domestic outbreaks, may help clarify whether virus circulation is decreasing or simply shifting further into wildlife reservoirs. Such approaches would support earlier detection of virus reintroduction or silent persistence and help guide more targeted control measures in domestic and wild populations

Supplementary Materials: The following supporting information can be downloaded at: https://www.mdpi.com/article/10.3390/vetsci12111086/s1; Table S1: Accession Numbers and Classification of Sequenced Samples.

Author Contributions: All authors contributed to the study's conception, design and article writing. D.G. and Š.G.I. were the principal investigators, and S.Š., I.T. and A.H.B. carried out the laboratory analysis. D.G. and Š.G.I. analyzed the results, while T.G. and V.M. supervised the study. All authors have read and agreed to the published version of the manuscript.

Funding: This study was funded by the Serbian Ministry of Science, Technological Development and Innovation (Contract No 451-03-136/2025-03/200030).

Institutional Review Board Statement: Ethical review and approval were waived for this study since the samples were obtained from pigs as part of the national monitoring for ASFV control.

Data Availability Statement: The original contributions presented in this study are included in the article/Supplementary Material. Further inquiries can be directed to the corresponding author.

Acknowledgments: The Serbian sequences were generated through the Sequencing Service of the Animal Production and Health Sub-Programme of the Joint FAO-IAEA Centre in Vienna, Austria.

Conflicts of Interest: The authors declare no conflicts of interest.

References

1. Alonso, C.; Borca, M.; Dixon, L.; Revilla, Y.; Rodriguez, F.; Escribano, J.M.; ICTV Report Consortium. ICTV Virus Taxonomy Profile: Asfarviridae. *J. Gen. Virol.* **2018**, *99*, 613–614. [CrossRef] [PubMed]
2. Blome, S.; Franzke, K.; Beer, M. African Swine Fever—A Review of Current Knowledge. *Virus Res.* **2020**, *287*, 198099. [CrossRef] [PubMed]
3. Goatley, L.C.; Freimanis, G.L.; Tennakoon, C.; Bastos, A.; Heath, L.; Netherton, C.L. African Swine Fever Virus NAM P1/95 Is a Mixture of Genotype I and Genotype VIII Viruses. *Microbiol. Resour. Announc.* **2024**, *13*, e0006724. [CrossRef] [PubMed]
4. Gallardo, C.; Casado, N.; Soler, A.; Djadjovski, I.; Krivko, L.; Madueño, E.; Nieto, R.; Perez, C.; Simon, A.; Ivanova, E.; et al. A Multi Gene-Approach Genotyping Method Identifies 24 Genetic Clusters within the Genotype II-European African Swine Fever Viruses Circulating from 2007 to 2022. *Front. Vet. Sci.* **2023**, *10*, 1112850. [CrossRef]
5. Milićević, V.; Kureljušić, B.; Maksimović Zorić, J.; Savić, B.; Stanojević, S.; Milakara, E. First Occurence of African Swine Fever in Serbia. *Acta Vet.* **2019**, *69*, 443–449. [CrossRef]
6. European Commission Animal Disease Information System (ADIS) 2023; Brussels, 2023. Available online: https://webgate.ec.europa.eu/tracesnt/adis/public/notification/outbreaks-overview-reports?reportYear=2023(accessed on 3 July 2025).
7. European Commission Animal Disease Information System (ADIS) 2024; Brussels, 2024. Available online:https://webgate.ec.europa.eu/tracesnt/adis/public/notification/outbreaks-overview-reports?reportYear=2024 (accessed on 3 July 2025).
8. European Commission Animal Disease Information System (ADIS) 2025; Brussels, 2025. Available online:https://webgate.ec.europa.eu/tracesnt/adis/public/notification/outbreaks-overview-reports?reportYear=2025 (accessed on 3 July 2025).
9. Glišić, D.; Milićević, V.; Veljović, L.; Milovanović, B.; Kureljušić, B.; Đorđević, I.; Anđelković, K.; Petković, J.; Dačić, M. Patterns of ASFV Transmission in Domestic Pigs in Serbia. *Pathogens* **2023**, *12*, 149. [CrossRef]
10. Glišić, D.; Šolaja, S.; Veljović, L.; Maksimović-Zorić, J.; Milićević, V. Spatiotemporal Analysis of African Swine Fever in Wild Boar in Serbia from 2020 to 2024. *Onderstepoort J. Vet. Res.* **2025**, *92*, 2209. [CrossRef]
11. Šolaja, S.; Glišić, D.; Milićević, V. Prevalence of Porcine Circoviruses 2 and 3 in Wild Boar in Serbia. *J. Vet. Diagn. Investig.* **2025**, *37*, 674–678. [CrossRef]
12. Iacolina, L.; Pertoldi, C.; Amills, M.; Kusza, S.; Megens, H.J.; Bâlteanu, V.A.; Bakan, J.; Cubric-Curic, V.; Oja, R.; Saarma, U.; et al. Hotspots of Recent Hybridization between Pigs and Wild Boars in Europe. *Sci. Rep.* **2018**, *8*, 17372. [CrossRef]
13. European Commission Pig Livestock Statistics. Available online: https://agriculture.ec.europa.eu/index_en?wt-search=yes (accessed on 3 July 2025).
14. Eurostat Pig Population—Annual Data. Available online: https://ec.europa.eu/eurostat/databrowser/view/apro_mt_lspig/default/table?lang=en (accessed on 3 July 2025).
15. Statistical Office of the Republic of Serbia Livestock Breeding. Available online: https://www.stat.gov.rs/en-us/oblasti/poljoprivreda-sumarstvo-i-ribarstvo/stocarstvo/ (accessed on 3 July 2025).
16. Glišić, D.; Milićević, V.; Krnjaić, D.; Toplak, I.; Prodanović, R.; Gallardo, C.; Radojičić, S. Genetic Analysis Reveals Multiple Intergenic Region and Central Variable Region in the African Swine Fever Virus Variants Circulating in Serbia. *Vet. Res. Commun.* **2023**, *47*, 1925–1936. [CrossRef] [PubMed]
17. Bastos, A.D.S.; Penrith, M.L.; Cruci Re, C.; Edrich, J.L.; Hutchings, G.; Roger, F.; Couacy-Hymann, E.; Thomson, G. Genotyping Field Strains of African Swine Fever Virus by Partial P72 Gene Characterisation. *Arch. Virol.* **2003**, *148*, 693–706. [CrossRef]
18. Forth, J.H.; Calvelage, S.; Fischer, M.; Hellert, J.; Sehl-Ewert, J.; Roszyk, H.; Deutschmann, P.; Reichold, A.; Lange, M.; Thulke, H.-H.; et al. African Swine Fever Virus—Variants on the Rise. *Emerg. Microbes Infect.* **2023**, *12*, 2146537. [CrossRef]
19. Cwynar, P.; Stojkov, J.; Wlazlak, K. African Swine Fever Status in Europe. *Viruses* **2019**, *11*, 310. [CrossRef]
20. Council of Ministers of Bosnia and Herzegovina. *Pravilnik o Mjerama Za Nadzor Afričke Svinjske Kuge*; Council of Ministers of Bosnia and Herzegovina: Sarajevo, Bosnia and Herzegovina, 2011.

21. Ministry of Agriculture, Forestry and Water Management. ПРАВИЛНИК о МерамаЗаРано Откривање, Дијагностику, Спречавање Ширења, Сузбијање и Искорењивање Заразне Болести Афричке Куге Свиња; Serbia, 2010. Available online: https://www.vet.minpolj.gov.rs/legislativa/pravilnici/Pravilnik_AKS_2010.pdf (accessed on 3 July 2025).
22. King, D.P.; Reid, S.M.; Hutchings, G.H.; Grierson, S.S.; Wilkinson, P.J.; Dixon, L.K.; Bastos, A.D.S.; Drew, T.W. Development of a TaqMan® PCR Assay with Internal Amplification Control for the Detection of African Swine Fever Virus. *J. Virol. Methods* **2003**, *107*, 53–61. [CrossRef]
23. WOAH. *Addressing African Swine Fever: Protocols and Guidelines for Laboratory Diagnosis*; WOAH: Paris, France, 2024.
24. Gallardo, C. African Swine Fever Virus P72 Genotype IX in Domestic Pigs, Congo, 2009. *Emerg. Infect. Dis.* **2011**, *17*, 1556–1558. [CrossRef] [PubMed]
25. Gallardo, C.; Fernández-Pinero, J.; Pelayo, V.; Gazaev, I.; Markowska-Daniel, I.; Pridotkas, G.; Nieto, R.; Fernández-Pacheco, P.; Bokhan, S.; Nevolko, O.; et al. Genetic Variation among African Swine Fever Genotype II Viruses, Eastern and Central Europe. *Emerg. Infect. Dis.* **2014**, *20*, 1544–1547. [CrossRef]
26. Mazur-Panasiuk, N.; Walczak, M.; Juszkiewicz, M.; Woźniakowski, G. The Spillover of African Swine Fever in Western Poland Revealed Its Estimated Origin on the Basis of O174L, K145R, MGF 505-5R and IGR I73R/I329L Genomic Sequences. *Viruses* **2020**, *12*, 1094. [CrossRef] [PubMed]
27. Okonechnikov, K.; Golosova, O.; Fursov, M. Unipro UGENE: A Unified Bioinformatics Toolkit. *Bioinformatics* **2012**, *28*, 1166–1167. [CrossRef] [PubMed]
28. Glišić, D.J. Izolacija i Genetska Karakterizacija Virusa Afričke Kuge Svinja u Srbiji u Periodu 2019–2023. Godine. Универзитет у Београду, 2025. Available online: https://nardus.mpn.gov.rs/handle/123456789/23333 (accessed on 30 September 2025).
29. Stončiūtė, E.; Malakauskas, A.; Conraths, F.J.; Masiulis, M.; Sauter-Louis, C.; Schulz, K. The Perceptions of Lithuanian Hunters towards African Swine Fever Using a Participatory Approach. *BMC Vet. Res.* **2022**, *18*, 401. [CrossRef]
30. Probst, C.; Globig, A.; Knoll, B.; Conraths, F.J.; Depner, K. Behaviour of Free Ranging Wild Boar towards Their Dead Fellows: Potential Implications for the Transmission of African Swine Fever. *R. Soc. Open Sci.* **2017**, *4*, 170054. [CrossRef]
31. Forth, J.H.; Forth, L.F.; Blome, S.; Höper, D.; Beer, M. African Swine Fever Whole-Genome Sequencing—Quantity Wanted but Quality Needed. *PLoS Pathog.* **2020**, *16*, e1008779. [CrossRef] [PubMed]

veterinary sciences

MDPI

Article

The Giant Panda Transferrin Receptor Facilitates Feline Parvovirus Infection to Drive Cross-Species Transmission

Qigui Yan [1,2,†], Huanyuan Hu [1,2,†], Shan Zhao [1,2], Qin Zhao [1], Rui Wu [1], Xiaobo Huang [1], Yiping Wang [1], Yiping Wen [1], Yi Zheng [1], Fei Zhao [1], Sanjie Cao [1], Senyan Du [1] and Yifei Lang [1,2,*]

[1] College of Veterinary Medicine, Sichuan Agricultural University, Chengdu 611130, China; yanqigui@126.com (Q.Y.)

[2] Chengdu National Agricultural Science and Technology Center, Chengdu 610213, China

* Correspondence: y_langviro@163.com

† These authors contributed equally to this work.

Simple Summary

Feline parvovirus (FPV) is a dangerous virus that causes severe illness in cats, including vomiting, diarrhea, and life-threatening infections. Recently, infection with this virus has also been reported in giant pandas, leading to similar symptoms and even death, putting these endangered animal species at risk. To understand how FPV spreads to pandas, we studied a panda-derived strain of the virus and examined how it interacts with panda host factors. The FPV virus uses a specific protein—namely the transferrin receptor 1—to enter and multiply inside cells. By testing this interaction in human cells that normally resist infection, we confirmed that the panda version of this receptor allows the virus to attach, enter, and replicate efficiently. These findings help explain why giant pandas are vulnerable to FPV and provide crucial insights for developing vaccines or treatments to protect them. Our research not only advances the understanding of cross-species virus transmission but also supports conservation efforts to safeguard giant pandas from deadly outbreaks.

Abstract

Feline parvovirus (FPV) causes feline panleukopenia, a highly contagious disease in cats, marked by severe leukopenia, biphasic fever, diarrhea, vomiting, and hemorrhagic enteritis. Recently, FPV infection in giant pandas has increased, causing diarrhea and ultimately fatal outcomes, thereby threatening their survival and reproduction. Here, we investigated the transmission of FPV in giant pandas and its interaction with cellular receptors using an FPV strain (pFPV-sc) isolated from giant panda feces. Recombinant feline transferrin receptor 1 (fTfR1) and the giant panda ortholog (gpTfR1) were expressed in non-susceptible HEK293T and HeLa cells, while viral infection levels were measured to determine the effect of gpTfR1 on pFPV-sc replication. The findings indicated that gpTfR1 overexpression in non-susceptible cells significantly enhanced pFPV-sc replication, particularly influencing the viral attachment and internalization stages. Our data further revealed early-stage colocalization between gpTfR1 expression and virus infection, suggesting that gpTfR1 facilitates early viral infection and replication. Taken together, our study provides the first evidence on the mechanism of FPV cross-species infection in giant pandas and elucidates the interaction between gpTfR1 and FPV, which establishes a theoretical basis for the development of preventive and therapeutic strategies, thereby safeguarding the health and survival of giant panda populations from FPV.

Keywords: feline parvovirus; transferrin receptor 1; virus replication; giant panda

1. Introduction

Feline parvovirus (FPV) causes a highly contagious acute infectious disease in cats named feline panleukopenia, with clinical symptoms characterized by severe leukopenia, biphasic fever, diarrhea, vomiting, and hemorrhagic enteritis [1]. FPV is characterized by its rapid evolution and spread. In addition to cats, cases of FPV cross-species transmission and infection in other animal species were frequent, such as deer, monkey, mink, tiger, leopard, lion, and panda, with isolates that retain over 95% of sequence homology across the viral genome, suggesting broad host adaptability despite conserved viral genetic architecture. Such infection often results in severe gastroenteritis, acute leukopenia, dehydration, lethargy, and high mortality, particularly in juveniles, due to the virus's tropism for rapidly dividing cells located in the intestinal epithelium and hematopoietic tissues [2–9]. In recent years, FPV infection in giant pandas has become increasingly common, leading to diarrhea and, in severe cases, mortality. This poses a potential threat to the survival and reproduction of giant pandas and a substantial challenge to the conservation of endangered wildlife.

FPV belongs to the *Parvoviridae* family and *Parvovirus* genus. It is a non-enveloped, single-stranded DNA virus whose capsid has icosahedral symmetry and a diameter of approximately 25 nm. The full-length viral genome is approximately 5.1 kb and contains two major open reading frames. Through alternative splicing, the nonstructural proteins (NS1 and NS2) and capsid proteins (VP1 and VP2) are encoded by the same mRNA [10]. VP2 is the main capsid protein of FPV and contains many overlapping antigenic and binding sites for the host transferrin type 1 receptor (TfR1) [11,12]. TfR1 plays a key role in the immune responses and determines the viral ligand for host specificity and tissue tropism [13]. Although the VP2 sequences of different parvoviruses are very similar, they exhibit high host specificity, where even substitutions in one or two amino acids can contribute to host tropism. Previously, we isolated an FPV strain, pFPV-sc, from a group of infected giant pandas with mortality cases. Sequence analysis indicated 99.8% similarity between pFPV-sc and the closest FPV sequence, but pFPV-sc contains unique characteristics in its VP2 protein [8].

Virus-receptor interaction is the initiating step in the process of virus infection and replication and plays a key regulatory role in the virus–host range, tissue tropism, and virus pathogenesis. Studies have shown that FPV infection in feline kidney cells is mediated by binding to the apical domain of feline transferrin receptor 1 (fTfR1) [14]. Notably, the TfR1 protein sequences of different hosts differ in their apical domains, which directly affect the host range of parvoviruses. For instance, during canine adaptation, contemporary canine parvovirus (CPV) variants evolved an enhanced binding affinity for canine TfR1, increasing infectivity in dogs while concurrently exhibiting reduced binding efficiency to feline TfR1 compared to ancestral CPV strains, reflecting host-specific adaptive selection pressures [15]. To date, only a few studies have shown that FPV can infect giant pandas; however, the mechanism by which FPV infects giant pandas and the transmission and evolution processes in new hosts remain largely undetermined. Therefore, comprehensive research into the mechanisms of infection is imperative. In this study, we analyzed the effect of giant panda transferrin receptor 1 (gpTfR1) on pFPV-sc infection. Our data indicated that specific binding occurs between FPV and the gpTfR1 located on the cell surface, thereby promoting the virus to invade host cells and infect giant pandas. This discovery lays the foundation for an in-depth understanding of the mechanisms underlying cross-host infection by feline parvoviruses.

2. Material and Methods

2.1. Cells, Virus, and Antibodies

The pFPV-sc strain of feline parvovirus from giant pandas was previously isolated, identified, and preserved in our laboratory [8]. The feline kidney (F81), human embryonic kidney (HEK293T), and human cervical cancer (HeLa) cell lines were originally purchased from ATCC (American Type Culture Collection, Manassas, VA, USA) and preserved in our laboratory.

The antibodies used in the study included Anti-V5 Tag mouse monoclonal antibody (ABT2170, Abbkine, Wuhan, China), Anti-Myc Tag mouse monoclonal antibody (ABT2060, Abbkine, Wuhan, China), DyLight 488, Goat anti-Mouse IgG (A23210, Abbkine, Wuhan, China), fluorescein isothiocyanate (FITC) Goat anti-Rabbit IgG H&L (ab6717, Abcam, Cambridge, UK), and Fluor647-conjugated goat anti-mouse IgG (A-21240, Life Technologies, Frederick, MD, USA).

2.2. Sequence Analysis and Plasmid Construction

Primers were designed for the CD region of F81 *TfR1* (GenBank accession number: NM001009312) (Table 1). Total RNA was extracted from the F81 cells. After reverse transcription, the obtained cDNA was used as a template for polymerase chain reaction (PCR) amplification and ligated into the pcDNA3.1-v5/myc vector. The PCR amplification conditions were as follows: 95 °C for 3 min; 35 cycles of 95 °C for 15 s, 58 °C for 30 s, 72 °C for 40 s; and 72 °C for 5 min; and storage at 4 °C. Concurrently, with reference to GenBank: XM_034660038.1, the CD region of *gpfR1* was synthesized by Sangon Biotech (Shanghai) Co., Ltd. and connected to the vector pcDNA3.1-v5/myc. The successfully constructed plasmids were named pcDNA3.1-v5/myc-fTfR and pcDNA3.1-v5/myc-gpTfR1. The homology between the *fTfR1* and *gpTfR1* sequences was analyzed using Clustal Omega software (version 1.2.2).

Table 1. Amplification primers and target fragments.

Primer	Sequence (5′-3′)	PCR Products/(bp)
F81-fTfR1-F	aaaaaGGATCCATGATGGATCAAGCCAGAT	2310
F81-fTfR1-R	aaaaaCTCGAGAAACTCATTGTCAATATCCCA	
TfR1-mRNA-F	TGGCTGTATTCTGCTCGTGG	189
TfR1-mRNA-R	GCCCCAAAAGATATGTCGG	
F81-RPL17-F	CTCTGGTCATTGAGCACATCC	123
F81-RPL17-R	TCAATGTGGCAGGGAGAGC	
293-βactin-F	CAAAAGGCGGGGTCGCAAT	108
293-βactin-R	CGACGATGGAAGGAAACACG	
Hela-PPIA-F	GAGGAAAACCGTGTACTATTAGC	86
Hela-PPIA-R	GGGACCTTGTCTGCAAAC	
VP2-qPCR-F	ACGGGTACTTTCAATAATCAGAC	179
VP2-qPCR-R	AATATCATCTAAAGCCATGTTTC	

Note: Primers F81-fTfR1-F and F81-fTfR1-R were used to construct the expression vector. Primers TfR1-mRNA-F, TfR1-mRNA-R, F81-RPL17-F, F81-RPL17-R, 293-βactin-F, 293-βactin-R, Hela-PPIA-F, and Hela-PPIA-R were used to detect the mRNA transcription level of TfR1. VP2-qPCR-F and VP2-qPCR-R were used to detect FPV copy number.

2.3. Indirect Immunofluorescence Detection of pFPV-sc Infection in Different Cells

When the F81, HEK293T, and HeLa cells in the 24-well plate grew to approximately 90% confluence, the cells were washed twice with phosphate-buffered saline (PBS), and the pFPV-sc (multiplicity of infection [MOI] = 0.1) virus solution was inoculated. After 48 h of infection, the cells were fixed with 4% paraformaldehyde, and an indirect immunofluorescence experiment was performed. First, the cells were washed twice with PBS, fixed with

4% paraformaldehyde for 30 min, and washed twice with PBS, while the cell membrane was permeabilized and then blocked with 2% bovine serum albumin for 2 h. Subsequently, the cells were incubated with a rabbit anti-FPV immune serum (1:200 dilution) for 1 h, as described previously [8]. After removing the primary antibody solution, the cells were washed twice with PBS and incubated with FITC goat anti-rabbit IgG antibody diluted 1:200 for 1 h. The incubation solution was discarded and the cells were washed twice with PBS. Next, 4′,6-diamidino-2-phenylindole (DAPI) diluted 1:100 was added and incubated at room temperature for 10 min. The cells were then washed twice with PBS. Finally, the cells were mounted with Prolong Diamond (P36971, Life Technologies, MD, USA) according to the manufacturer's instructions. The cells were observed and photographed using a fluorescence microscope.

2.4. qPCR Verification of TfR1 Overexpression and FPV Copy Number

HEK293T and HeLa cells were seeded in 12-well plates and incubated until 70% confluence for transient transfection. Next, using the Lipofectamine™ 3000 (Life Technologies, USA) transfection method, 1 μg of pcDNA3.1-v5/myc-fTfR1 and pcDNA3.1-v5/myc-gpTfR1 plasmids was transfected. Concurrently, pcDNA3.1-v5/myc was used as a negative control (pcDNA3.1) and the untransfected blank group was used as a blank control (mock). Total RNA was extracted using the TRIzol reagent and reverse-transcribed. The obtained cDNA was used to detect overexpression of the recombinant plasmids using quantitative PCR (qPCR). The reaction mixture consisted of 2 × TB Green Premix Ex TaqTM II (RR820Q, Takara Biomedical Technology, Dalian, China) (10 μL); 0.5 pmol of upstream and downstream primers; and 7 μL of ddH_2O; and 1 μL of cDNA. The reaction conditions were as follows: 95 °C for 30 s, 40 cycles of 95 °C for 5 s, and 60 °C for 30 s. The qPCR results were expressed using the $2^{-\Delta\Delta CT}$ method, with β-actin serving as the endogenous control for normalization. The threshold cycle (CT) values of TfR1 were normalized to β-actin ($\Delta CT = CT_{TfR1} - CT_{\beta\text{-actin}}$), followed by calibration to a reference sample ($\Delta\Delta CT = \Delta CT_{sample} - \Delta CT_{calibrator}$). Relative TfR1 expression was calculated as $2^{-\Delta\Delta CT}$.

After confirming that the overexpression was effective, and after infection with pFPV-sc (MOI = 0.1) for 36 h, the DNA of the viral solution was extracted, and the FPV copy number was determined by absolute quantification using a standard curve generated from serial 10-fold dilutions (10^7–10^1 copies/μL) of a linearized plasmid containing the FPV VP2 gene (Table 1). Cq values obtained from samples were interpolated from the standard curve, with copy numbers calculated using the formula [$6.022 \times 10^{23} \times$ (concentration in ng/μL) $\times 10^9$]/(plasmid length × 660).

2.5. Indirect Immunofluorescence Verification of TfR1 Overexpression and pFPV-sc Replication Levels

HEK293T and HeLa cells were plated in 24-well plates containing circular coverslips. Next, 0.5 μg of pcDNA3.1-v5/myc-fTfR1 and pcDNA3.1-v5/myc-gpTfR1 plasmids were transfected. Concurrently, pcDNA3.1-v5/myc was used as a negative control (pcDNA3.1), and the untransfected blank group was used as a blank control (mock). After 36 h of transfection, the overexpression of *fTfR1* and *gpTfR1* was detected using an indirect fluorescence immunoassay. V5 and Myc-tagged mouse monoclonal antibodies were used as the primary antibodies, and DyLight 488 goat anti-mouse IgG was used as the secondary antibody. Finally, fluorescence was observed using a fluorescence microscope, and the localization of TfR1 was analyzed.

Concurrently, cells were inoculated with pFPV-sc (MOI = 0.1) for 36 h and fixed with 4% paraformaldehyde. Rabbit anti-pFPV-sc immune serum was used as the primary antibody, and FITC goat anti-rabbit IgG was used as the secondary antibody. Finally, fluorescence was observed using a fluorescence microscope to determine the number of FPV-infected cells.

2.6. Effect of gpTfR1 Expression on pFPV-sc Adsorption and Internalization

HEK293T and HeLa cells were inoculated with pFPV-sc (MOI = 0.5), incubated at 4 °C for 1 h, and washed twice with cold PBS to remove unadsorbed viruses. Next, three freeze–thaw treatments were performed to release the virus particles adsorbed on the cell surface. The supernatant was collected, viral DNA was extracted, and the viral Cq value in the adsorption stage was obtained using qPCR to calculate the FPV copy number. Concurrently, in a 24-well plate with a circular coverslip, the above procedure was repeated, the cells were fixed with 4% paraformaldehyde, and an indirect fluorescence immunoassay was performed to detect the viral fluorescence signals during the adsorption stage.

HEK293T and HeLa cells were inoculated with pFPV-sc (MOI = 0.5), incubated at 4 °C for 1 h, and then transferred to 37 °C and incubated for 1 h. The virus solution was removed and the unadsorbed viruses were removed by washing with pre-cooled PBS. Proteinase K solution (400 µL, 1 mg/mL) was added to each well to remove viruses bound to the cell surface. Three freeze–thaw treatments were performed to extract viral DNA and obtain the viral Cq value in the internalization stage using qPCR to calculate the FPV copy number. Concurrently, in a 24-well plate with a circular coverslip, after allowing internalization at 37 °C, the cells were immediately fixed with 4% paraformaldehyde at room temperature for 30 min. An indirect fluorescence immunoassay was performed to detect viral fluorescence signals during internalization. This procedure required the permeabilization of the cell membrane.

2.7. Colocalization Analysis of gpTfR1 Expression and pFPV-sc Infection

HEK293T and HeLa cells transfected with the recombinant plasmids were infected with pFPV-sc (MOI = 0.1) at 37 °C for 1 h. The cells were then thoroughly washed with PBS and fixed with 4% paraformaldehyde. Rabbit anti-FPV immune serum and the V5-tagged mouse monoclonal antibody were used as primary antibodies, and FITC-conjugated goat anti-rabbit IgG (green light) and Fluor647-conjugated goat anti-mouse IgG (red light) were used as secondary antibodies. Finally, the fluorescence signals were observed using a fluorescence microscope.

2.8. Examination of Cell Viability Detection by CCK8 Assay

The cell suspension was inoculated into a 96-well plate and incubated for 16 h. Plasmid transfection was performed for 36 h. After transfection, the cell supernatants were discarded. Next, 10 µL of the CCK8 solution and 90 µL of 10% Dulbecco's modified Eagle's medium were added to each well. After culturing for 2 h, the absorbance was measured at 450 nm using a microplate reader. The formula used is as follows: Cell viability (%) = (As − Ab)/(Ac − Ab) × 100%, where As is the OD value of the transfection experimental well, Ab is the OD value of the pcDNA3.1 empty vector-transfected well, and Ac is the OD value of the normal cell control well.

2.9. Data Analysis

All experiments were repeated at least three times. In the indirect immunofluorescence experiments, ImageJ software (version 1.52e) was used to count the infected cells. All statistical analyses were performed using unpaired two-tailed tests in GraphPad Prism version 9. Data are presented as the standard error of the mean of three independent experiments or replicates. Significance levels are as follows: ns, not significant; * $p < 0.05$; ** $p < 0.01$; and *** $p < 0.001$. Homology and amino acid site homology analyses of *fTfR1*

(GenBank: NM001009312) and *gpTfR1* (GenBank: XM044382865) were performed using Clustal Omega software (version 1.2.2).

3. Results

3.1. Validation of Cell Tropism of the Giant Panda Derived FPV (pFPV-sc)

Building upon previous observations, F81, HEK293T, and HeLa cells were selected for analysis of the cell tropism of pFPV-sc and the possibility of transducing exogenous TfR1 expression that might support pFPV-sc infection. Cells were inoculated with pFPV-sc at a multiplicity of infection [MOI] of 0.1, and indirect immunofluorescence was used to evaluate infection levels at 48 h post-infection. As shown in Figure 1, fluorescence signals were observed in F81 cells, whereas only a few fluorescence signals were observed in HEK293T and HeLa cells. This result indicates that pFPV-sc showed high susceptibility in F81 cells, whereas infectivity was relatively low in HEK293T and HeLa cells, which could be applied for examination for the effect of foreign TfR1 expression on virus infection.

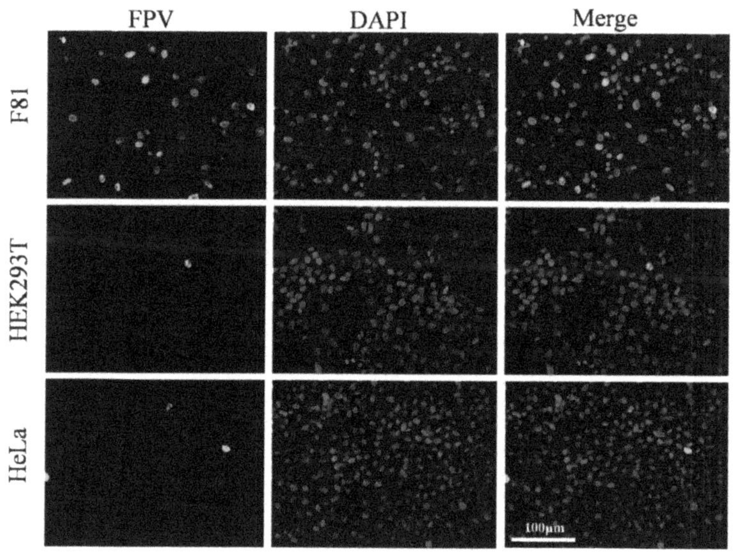

Figure 1. pFPV-sc infection in F81, HEK293T, and HeLa cells. Green fluorescence represents successful pFPV-sc infection.

3.2. Conformation of Expression of Recombinant *fTfR1* and *gpTfR1*

HEK293T and HeLa cells were transfected with pcDNA3.1-v5/myc-fTfR1 and pcDNA3.1-myc/v5-gpTfR1 recombinant plasmids, with pcDNA3.1-v5/myc empty plasmids as controls. The cell viability in each group was determined using the CCK8 method. The results showed that although cell viability in the cells transfected with *fTfR1* and *gpTfR1* slightly decreased, no significant difference compared to that of the mock and empty vector groups. Total RNA was extracted from each transfection group, and the *TfR1* mRNA content in each group of cells was measured. The results indicate that the *TfR1* mRNA content in the groups transfected with *fTfR1* and *gpTfR1* was significantly increased. In addition, using an indirect immunofluorescence assay, obvious fluorescence signals were observed in cells transfected with *fTfR1* and *gpTfR1*, whereas no fluorescence signal was observed in the control group (Figure 2). These results indicate that recombinant *fTfR1* and *gpTfR1* can be successfully induced in non-susceptible cells of pFPV-sc without altering cell viability.

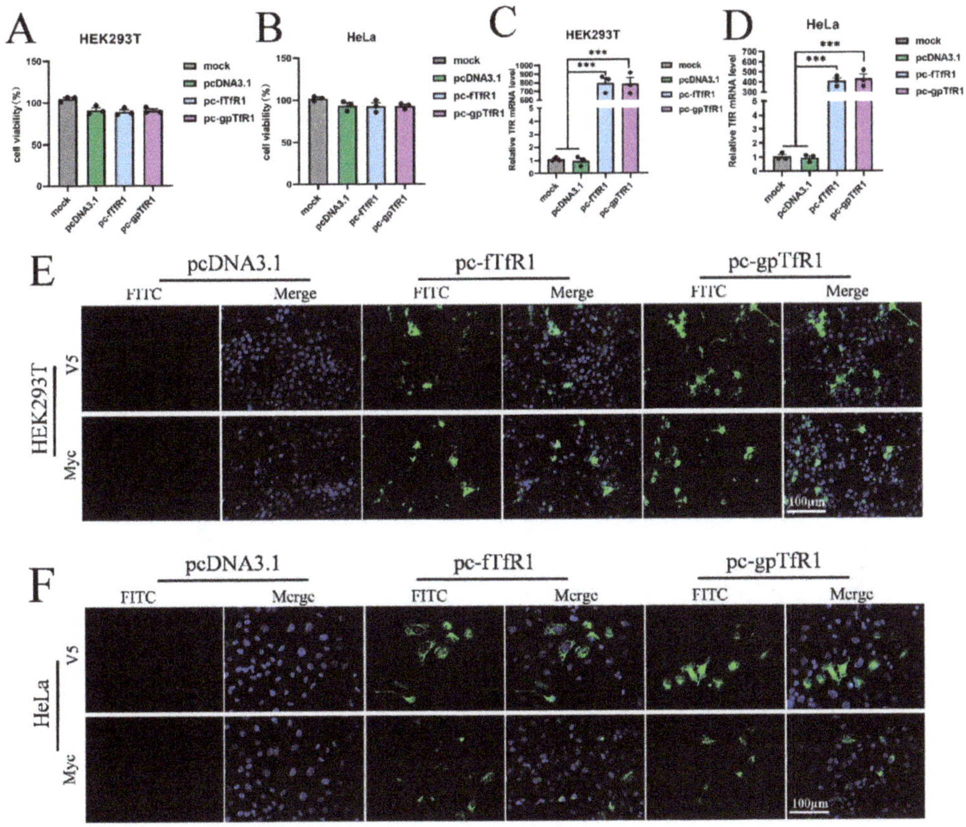

Figure 2. Expression identification of the recombinant fTfR1 and gpTfR1 plasmids. (**A**,**B**) CCK8 assay to detect cell viability after transfection with recombinant plasmids. (**C**,**D**) qPCR to detect the mRNA level of TfR1 in cells. (**E**,**F**) Indirect immunofluorescence assay against V5 or Myc tag to detect the expression of the fTfR1 and gpTfR1 overexpression vectors. (Blue: nucleus; Green: V5 or Myc tag; *** $p < 0.001$).

3.3. Giant Panda TfR1 Promotes pFPV-sc Replication

We next examined the probable effect of recombinant gpTfR1 on pFPV-sc infection in non-susceptible cells. Recombinant gpTfR1 or fTfR1 were expressed in HEK293T and HeLa cells prior to virus infection, and the results of qPCR analysis revealed that the viral copy number in the gpTfR1 group increased significantly in both cell lines in comparison with mock-transfected or empty vector-transfected cells (Figure 3A,B). Similar observations were obtained with indirect immunofluorescence analysis, where a strong fluorescence signal was observed in the gpTfR1 transfected group and the infection rate reached above 20% (Figure 3C–F). Differences were still observed in both the viral copy number and cell infection rate in the gpTfR1 group in comparison with the fTfR1 group, indicating a relatively lower viral replication (Figure 3). This may be due to the difference in amino acid sites between the two TfR1 receptors, resulting in different levels of receptor–virus interaction, where pFPV-sc still prefers fTfR1 over gpTfR1. Apparently, the expression of gpTfR1 is already favored by pFPV-sc, resulting in constructive infection in FPV-unsusceptible cells.

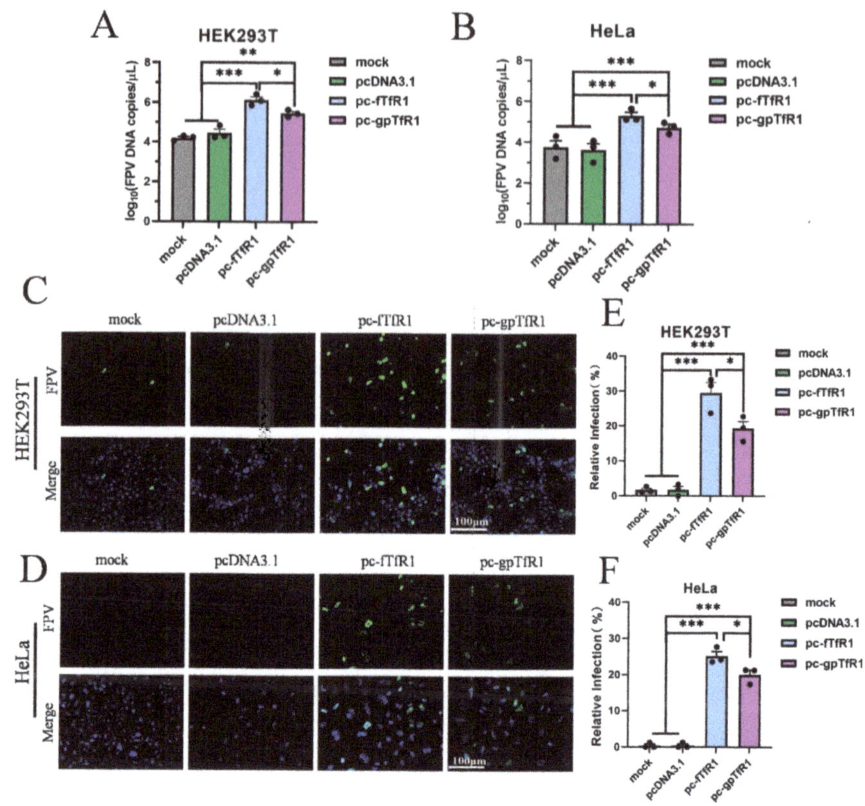

Figure 3. Overexpression of gpTfR1 promotes pFPV-sc replication. (**A**,**B**) qPCR to detect FPV copies. (**C**,**D**) Indirect immunofluorescence assay to detect the number of infected cells (Blue: nucleus; Green: FPV). (**E**,**F**) ImageJ counts of the percentage of infected cells. (* $p < 0.05$, ** $p < 0.01$, *** $p < 0.001$).

3.4. Overexpression of gpTfR1 Promotes pFPV-sc Adsorption and Internalization

After confirming the influence of gpTfR1 on viral replication, the effect of gpTfR1 on the entry of pFPV-sc into cells was further studied. By measuring viral copy numbers and performing indirect immunofluorescence experiments, we determined whether gpTfR1 expression affects specific stages of pFPV-sc binding or endocytosis. The results showed that viral copy numbers in both fTfR1 and gpTfR1 groups increased significantly during the virus adsorption stage. Concurrently, compared with the fTfR1 group, the viral copy number in the gpTfR1 group was slightly lower (Figure 4A,C). In the meantime, obvious fluorescence signals were observed in the fTfR1 and gpTfR1 groups via indirect fluorescence immunoassay, whereas no fluorescence signal was observed in the control groups (Figure 4B,D). At the viral internalization stage, the viral copy numbers in the fTfR1 and gpTfR1 groups increased significantly (Figure 4E,G), while no significant difference was observed between the fTfR1 and gpTfR1 groups. Contemporarily, through the detection of fluorescence signals via an indirect fluorescence immunoassay, obvious fluorescence signals were observed in the fTfR1 and gpTfR1 groups, whereas minimal or no fluorescence signal was observed in the control groups (Figure 4F,H). These results indicate that the expression of fTfR1 and gpTfR1 enhances virus attachment and internalization, and that gpTfR1 has a lower effect on increasing virus copy numbers at the virus adsorption stage than fTfR1.

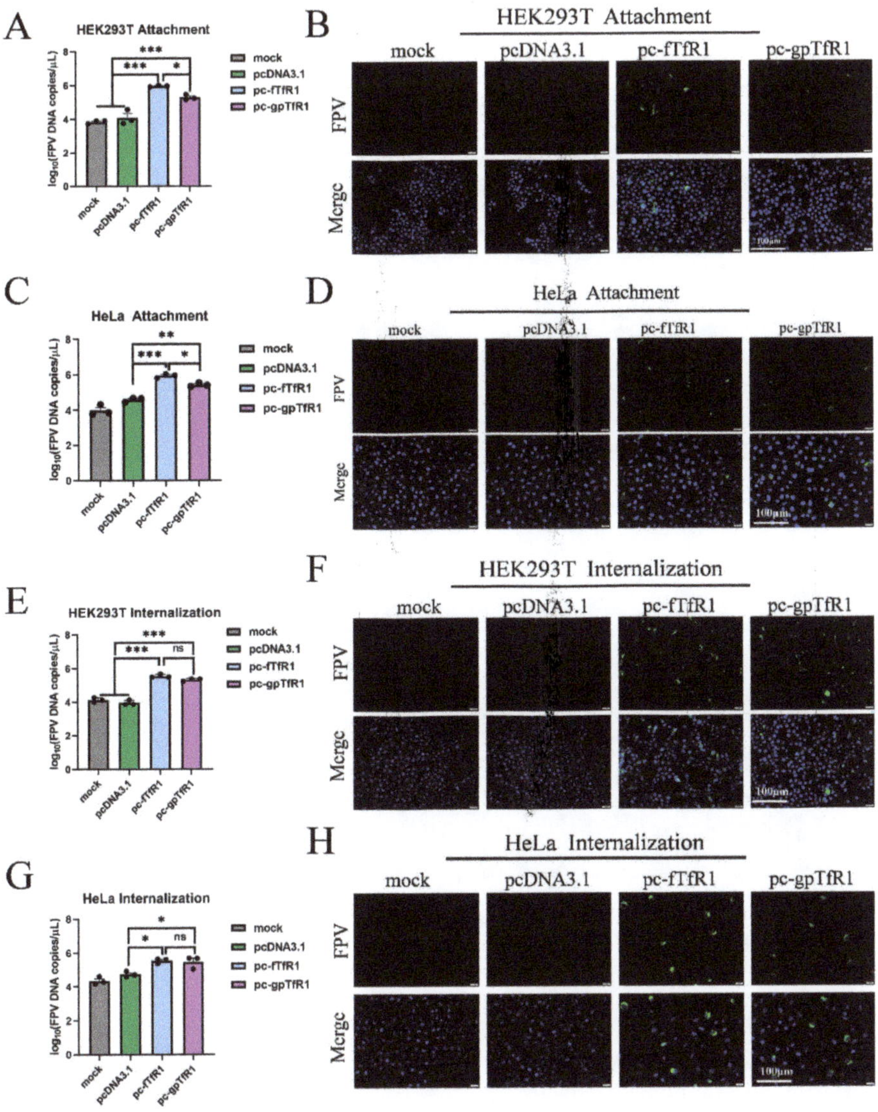

Figure 4. Effect of gpTfR1 expression on the internalization stage of pFPV-sc adsorption. (**A**,**C**,**E**,**G**) FPV copy numbers detected using qPCR. (**B**,**D**,**F**,**H**) The number of infected cells detected by indirect immunofluorescence assay. (Blue: nucleus; Green: FPV; ns: not significant; * $p < 0.05$, ** $p < 0.01$, *** $p < 0.001$).

3.5. pFPV-sc Infection in Unsusceptible Cells Requires gpTfR1 Expression

To further validate that pFPV-sc infection and gpTfR1 expression occur in the same cell post-transfection, the subcellular localization of gpTfR1 and pFPV-sc was analyzed during the early stages of pFPV-sc infection. The results showed that virus infection (green) primarily colocalized with recombinant gpTfR1 (red), indicating that an early virus infection stage correlates with gpTfR1 expression (Figure 5). A similar pattern was also found with fTfR1 expression. Taken together, our observation demonstrated that the gpTfR1 association plays an important role in the process of pFPV-sc infection, which could serve as an initial and necessary step for FPV cross-species transmission to the giant panda population.

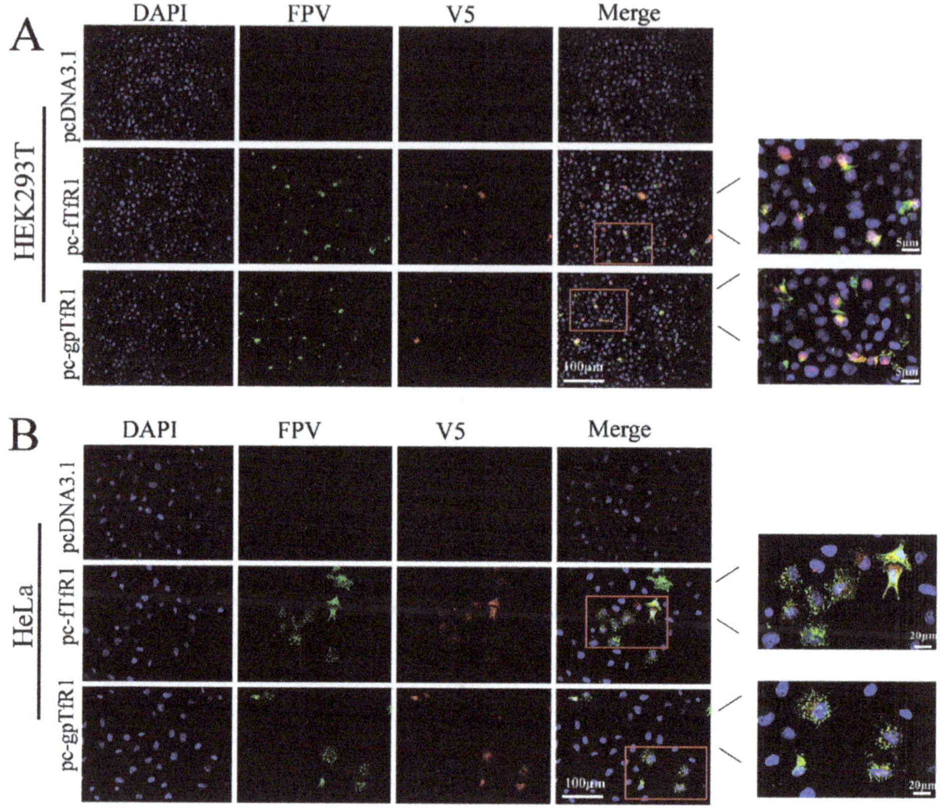

Figure 5. Immunofluorescence colocalization of gpTfR1 with pFPV-sc. (**A**) Colocalization of gpTfR1 and FPV in HEK293 cells. (**B**) Colocalization of gpTfR1 and FPV in HeLa cells. (Blue: nucleus; Green: FPV; Red: V5 tag).

4. Discussion

Most virus infections begin by binding to specific cell receptors and the uptake of the virus into cells through receptor-mediated endocytosis [16]. This process is controlled by multiple factors, including the characteristics of virus-binding receptors, signal transduction, and endocytic properties, as well as the affinity of the virus for the receptor and the structural characteristics of interactions under certain conditions. Parvoviruses bind TfR1 to enter cells through clathrin-mediated endocytosis, which is then transported via the endosomal pathway [17,18]. Studies have shown that certain viruses can bind to TfR1 with different binding levels and affinities within different hosts [19,20]. For example, although the expression levels of TfR1 in feline and canine cells are similar, the binding and endocytosis levels of feline cells to canine parvovirus (CPV)-2 and CPV-2b capsids are 3.5 times and 5 times higher than those of canine cells, respectively, which might be due to the relatively lower binding affinity of CPV capsids for canine cells [17]. At 37 °C, there is a difference in the binding levels of feline and canine TfR1 to CPV-2 capsids, and cells expressing fTfR1 show a higher level of binding and uptake. In the present study, we showed that expression of fTfR1 and gpTfR1 had a significant promoting effect on viral infection at both the virus adsorption and internalization stages. In addition, at the virus adsorption stage, compared to fTfR1, gpTfR1 exhibited a lower infection level, but the internalization levels were similar. The difference in replication levels between fTfR1 and gpTfR1 is likely due to the higher affinity of the virus against fTfR1, where the probable differences in the apical domains between fTfR1 and gpTfR1 may play an essential role, as with the host specificities of CPV and FPV [19,20]. Concurrently, the colocalization of gpTfR1 and pFPV-sc infection was analyzed by immunofluorescence colocalization experiments in the early stage of infection, and it was found that there was an aggregation of pFPV-sc at the location of gpTfR1. In conclusion, gpTfR1 may first promote infection by virus association, followed by promoting viral replication at the initial stage of viral infection.

Parvoviruses display restricted host specificity. For instance, studies have shown that FPV cannot bind to canine TfR1, which is only 12% different from fTfR1, indicating strong host specificity [12]. The amino acid homology between fTfR1 and gpTfR1 is approximately 87.66% [8]. FPV should also have a host barrier against gpTfR1. However, in natural ecosystems, the phenomenon of FPV cross-host infection of giant pandas has been observed. Therefore, using fTfR1 as a positive control, we explored the receptor for FPV infection in giant pandas. The results of this study showed that gpTfR1 expression in non-susceptible cells has a function similar to that of fTfR1 and can significantly promote the replication of pFPV-sc. This effect may have occurred at an early stage of viral infection, indicating that gpTfR1 may play an important role in the life cycle of FPV, possibly by promoting viral entry into cells. Although fTfR1 may be similar to gpTfR1 in some respects, there are still differences between them, and further research is needed to verify the exact role of gpTfR1 in FPV infection. FPV, which initially only infected felines, has evolved to cause a CPV pandemic in canines as a host-range variant through several key amino acid changes in the VP2 capsid protein [19]. Currently, they are no longer limited to felines and canines but can infect a wider range of animal groups, including Ursidae and primates. This trend has prompted consideration regarding the continuous evolution of FPV and raised questions about its potential to infect humans through mutual adaptation to the host TfR1. The present study found that pFPV-sc caused sporadic infections in human HEK293T and HeLa cells, suggesting a potential risk of cross-species transmission of FPV. Future research is required to further explore the evolutionary path of FPV and its interaction with the host to comprehensively assess its potential impact on human health and implement corresponding prevention and control measures.

In conclusion, gpTfR1 was used in the present study as a hypothetical surface receptor for virus–host binding, and the mechanism of FPV cross-species infection in giant pandas was investigated. These results showed that gpTfR1 is the cellular receptor for FPV infection in giant pandas and can promote FPV replication. This discovery is expected to provide a theoretical basis for the future development of antiviral intervention strategies and drugs that can prevent FPV infection from giant pandas.

Author Contributions: Conceptualization, Q.Y. and Y.L.; methodology, H.H. and Y.L.; software, H.H. and S.Z.; validation, S.Z., Q.Z., R.W., S.C. and Y.L.; formal analysis, X.H., Y.W. (Yiping Wang) and Y.W. (Yiping Wen); investigation, Q.Y. and H.H.; resources, Y.Z., S.Z. and S.D.; data curation, H.H.; writing—original draft preparation, Q.Y., H.H. and Y.L.; writing—review and editing, Y.L.; visualization, F.Z.; supervision, Q.Y. and Y.L.; project administration, Y.L.; funding acquisition, Q.Y. and Y.L. All authors have read and agreed to the published version of the manuscript.

Funding: This work was supported by the Joint Funds of Science Technology and Education of Sichuan Province (Grant No. 2024NSFSC2063), the Chengdu Giant Panda Breeding Research Foundation, China (Grant No. CPF2017-35) and the Self-supporting Project of Chengdu Giant Panda Breeding research Base, China (Grant No. 2021CPB-C13).

Institutional Review Board Statement: Not applicable.

Informed Consent Statement: Not applicable.

Data Availability Statement: All data shown in this manuscript is assessable upon reasonable request.

Conflicts of Interest: The authors declare no conflict of interest.

References

1. Jager, M.C.; Tomlinson, J.E.; Lopez-Astacio, R.A.; Parrish, C.R.; Van de Walle, G.R. Small but mighty: Old and new parvoviruses of veterinary significance. *Virol. J.* **2021**, *18*, 210. [CrossRef] [PubMed]
2. Wang, K.; Du, S.S.; Wang, Y.Q.; Wang, S.Y.; Luo, X.Q.; Zhang, Y.Y.; Liu, C.F.; Wang, H.J.; Pei, Z.H.; Hu, G.X. Isolation and identification of tiger parvovirus in captive siberian tigers and phylogenetic analysis of VP2 gene. *Infect. Genet. Evol.* **2019**, *75*, 103957. [CrossRef] [PubMed]
3. Studdert, M.; Kelly, C.; Harrigan, K. Isolation of panleucopaenia virus from lions. *Vet. Rec.* **1973**, *93*, 156–158. [CrossRef] [PubMed]
4. Huang, S.; Li, X.; Xie, W.; Guo, L.; You, D.; Xu, H.; Liu, D.; Wang, Y.; Hou, Z.; Zeng, X.; et al. Molecular detection of parvovirus in captive Siberian tigers and lions in Northeastern China from 2019 to 2021. *Front. Microbiol.* **2022**, *13*, 898184. [CrossRef] [PubMed]
5. Barker, I.K.; Povey, R.C.; Voigt, D.R. Response of mink, skunk, red fox and raccoon to inoculation with mink virus enteritis, feline panleukopenia and canine parvovirus and prevalence of antibody to parvovirus in wild carnivores in Ontario. *Can. J. Comp. Med.* **1983**, *47*, 188. [PubMed]
6. Yang, S.T.; Wang, S.J.; Feng, H.; Zeng, L.; Xia, Z.P.; Zhang, R.Z.; Zou, X.H.; Wang, C.Y.; Liu, Q.A.; Xia, X.Z. Isolation and characterization of feline panleukopenia virus from a diarrheic monkey. *Vet. Microbiol.* **2010**, *143*, 155–159. [CrossRef] [PubMed]
7. Yi, S.; Liu, S.; Meng, X.; Huang, P.; Cao, Z.; Jin, H.; Wang, J.; Hu, G.; Lan, J.; Zhang, D.; et al. Feline panleukopenia virus with G299E substitution in the VP2 protein first identified from a captive giant panda in China. *Front. Cell. Infect. Microbiol.* **2022**, *11*, 820144. [CrossRef] [PubMed]
8. Zhao, S.; Hu, H.Y.; Lan, J.C.; Yang, Z.S.; Peng, Q.L.; Yan, L.H.; Luo, L.; Wu, L.; Lang, Y.F.; Yan, Q.G. Characterization of a fatal feline panleukopenia virus derived from giant panda with broad cell tropism and zoonotic potential. *Front. Immunol.* **2023**, *14*, 1237630. [CrossRef] [PubMed]
9. Li, L.; Woods, L.; Gerstenberg, G.; Deng, X.; Delwart, E. A Common Parvovirus in Deer from California, USA. *J. Wildl. Dis.* **2016**, *52*, 962–964. [CrossRef] [PubMed]
10. Cotmore, S.F.; Agbandje-McKenna, M.; Chiorini, J.A.; Mukha, D.V.; Pintel, D.J.; Qiu, J.; Soderlund-Venermo, M.; Tattersall, P.; Tijssen, P.; Gatherer, D.; et al. The family Parvoviridae. *Arch. Virol.* **2013**, *159*, 1239–1247. [CrossRef] [PubMed]
11. Lee, H.; Callaway, H.M.; Cifuente, J.O.; Bator, C.M.; Parrish, C.R.; Hafenstein, S.L. Transferrin receptor binds virus capsid with dynamic motion. *Proc. Natl. Acad. Sci. USA* **2019**, *116*, 20462–20471. [CrossRef] [PubMed]
12. Hueffer, K.; Govindasamy, L.; Agbandje-McKenna, M.; Parrish, C.R. Combinations of two capsid regions controlling canine host range determine Canine transferrin receptor binding by canine and feline parvoviruses. *J. Virol.* **2003**, *77*, 10099–10105. [CrossRef] [PubMed]
13. Callaway, H.M.; Welsch, K.; Weichert, W.; Allison, A.B.; Hafenstein, S.L.; Huang, K.; Iketani, S.; Parrish, C.R. Complex and dynamic interactions between parvovirus capsids, transferrin receptors, and antibodies control cell infection and host range. *J. Virol.* **2018**, *92*, e00460-18. [CrossRef] [PubMed]
14. Harbison, C.E.; Chiorini, J.A.; Parrish, C.R. The parvovirus capsid odyssey: From the cell surface to the nucleus. *Trends Microbiol.* **2008**, *16*, 208–214. [CrossRef] [PubMed]
15. Hueffer, K.; Parker, J.; Weichert, W.S.; Geisel, R.E.; Sgro, J.Y.; Parrish, C.R. The natural host range shift and subsequent evolution of canine parvovirus resulted from virus-specific binding to the canine transferrin receptor. *J. Virol.* **2003**, *77*, 1718–1726. [CrossRef] [PubMed]
16. Chi, P.-I.; Liu, H.-J. Molecular signaling and cellular pathways for virus entry. *ISRN Virol.* **2013**, *2013*, 306595. [CrossRef]
17. Harbison, C.E.; Lyi, S.M.; Weichert, W.S.; Parrish, C.R. Early steps in cell infection by parvoviruses: Host-specific differences in cell receptor binding but similar endosomal trafficking. *J. Virol.* **2009**, *83*, 10504–10514. [CrossRef] [PubMed]
18. Hueffer, K.; Parrish, C.R. Parvovirus host range, cell tropism and evolution. *Curr. Opin. Microbiol.* **2003**, *6*, 392–398. [CrossRef] [PubMed]
19. Cureton, D.K.; Harbison, C.E.; Cocucci, E.; Parrish, C.R.; Kirchhausen, T. Limited transferrin receptor clustering allows rapid diffusion of canine parvovirus into clathrin endocytic structures. *J. Virol.* **2012**, *86*, 5330–5340. [CrossRef] [PubMed]
20. Palermo, L.M.; Hueffer, K.; Parrish, C.R. Residues in the apical domain of the feline and canine transferrin receptors control host-specific binding and cell infection of canine and feline parvoviruses. *J. Virol.* **2003**, *77*, 8915–8923. [CrossRef] [PubMed]

veterinary sciences

MDPI

Article

Establishment of a One–Pot RAA–CRISPR/Cas13a Assay-Based TGEV S Gene Detection

Lindan Lv [1,2], Hao Mu [1,2,3,4], Shaomei Li [1,2,3,4], Jieqi Gao [1,2], Mingni Liu [1], Shuizhu Niu [1], Guoyang Xu [1,2,3,4], Lizhi Fu [1,2,3,4], Zhenhui Song [2,*] and Liu Yang [1,2,3,4,*]

1 National Center of Technology Innovation for Pigs, Chongqing 402460, China; lvlindan@foxmail.com (L.L.); mouh@cqaa.cn (H.M.); shaomeili123@163.com (S.L.); 15035607783@163.com (J.G.); declanlmnyui@163.com (M.L.); cecilianiu@outlook.com (S.N.); guoyangxu@126.com (G.X.); fulz@cqaa.cn (L.F.)
2 College of Veterinary Medicine, Southwest University, Chongqing 400715, China
3 Chongqing Academy of Animal Sciences, Chongqing 402460, China
4 Chongqing Research Center of Veterinary Biologicals Engineering and Technology, Chongqing 400715, China
* Correspondence: szh7678@swu.edu.cn (Z.S.); yangl@cqaa.cn (L.Y.)

Simple Summary: Porcine transmissible gastroenteritis virus (TGEV) is a major cause of severe diarrhea in pigs, leading to high mortality and significant economic losses. Due to the high genomic similarity between TGEV and porcine respiratory coronavirus (PRCV), it is difficult to distinguish them serologically. In this study, we developed a specific, rapid, and sensitive nucleic acid detection method for TGEV using a combination of recombinase–aided amplification (RAA) and the CRISPR/Cas13a system. This one–pot assay, performed at 37 °C, can detect TGEV within 40 min, with a sensitivity of 4.13 copies/μL. When tested on 140 clinical samples, the method demonstrated 100% accuracy compared with RT–qPCR and RT–PCR. Its simplicity and efficiency make it a promising tool for timely detection and control of TGEV infection.

Abstract: Porcine transmissible gastroenteritis virus (TGEV) is a highly contagious pathogen causing severe diarrhea in pigs, particularly piglets, leading to significant economic losses. Distinguishing TGEV from the genetically similar porcine respiratory coronavirus (PRCV) remains challenging due to their high genomic homology. In this study, we developed a one–pot assay combining recombinase-aided amplification (RAA) and CRISPR/Cas13a technology, targeting the TGEV S gene. This method was optimized for sensitivity and specificity, with orthogonal tests determining the optimal reagent concentrations. The assay achieved a detection limit of 4.13 copies/μL within 40 min at 37 °C, demonstrating no cross-reactivity with other porcine viruses. Clinical validation on 140 samples showed 100% concordance with RT–qPCR and RT–PCR results. Since the established method is completed in a single reaction tube, it eliminates the need for step-by-step operations, simplifying the process and reducing the risk of cross–contamination and false positives in subsequent tests. Overall, this assay shows promising potential for TGEV detection.

Keywords: TGEV; RAA; CRISPR/Cas13a; detection method

1. Introduction

Porcine transmissible gastroenteritis virus (TGEV) is a highly contagious coronavirus that causes severe diarrhea, vomiting, and dehydration in pigs. In piglets under two weeks of age, mortality rates can reach 100%, while surviving animals exhibit stunted growth, significantly reducing farm productivity and leading to significant economic losses in the swine industry due to high mortality, reduced growth rates, and costly containment measures [1–3]. There are no specific drugs or effective treatments for TGE, except for immunization. However, vaccines do not provide immune protection against TGEV variants. Therefore, timely detection and accurate diagnosis of TGEV infection are crucial for the scientific and effective prevention and control of the disease [4]. Traditional methods for TGEV diagnosis include virus isolation and identification, the ELISA, and PCR, among others. However, virus isolation is time-consuming, typically taking 3–7 days, and requires specialized biosecurity facilities. ELISA offers moderate sensitivity, with a detection limit of approximately 10^3 $TCID_{50}$/mL; its utility is limited by the high genetic and antigenic similarity between TGEV and PRCV [5–7], a TGEV S gene deletion mutant strain that shares up to 96% nucleic acid identity at the genomic level [8,9]. This makes it challenging to distinguish between the two viruses using serological methods. Although PCR and real–time PCR provide high sensitivity, with detection limits as low as 10^1–10^2 copies/μL, these methods mainly target conserved regions such as the M or N genes, making it difficult to distinguish between TGEV and PRCV in clinical settings.

In recent years, clustered regularly interspaced short palindromic repeat (CRISPR)/CRISPR–associated protein (Cas) systems have attracted significant scientific attention. Among them, the CRISPR/Cas13a detection platform has emerged as a powerful tool for molecular diagnostics, enabling the selective activation of Cas13a nuclease activity through guide RNA–mediated targeting of specific nucleic acids [10–12]. This technology has been successfully applied to detect various pathogens, including African swine fever virus (ASFV) [13], porcine epidemic diarrhea virus (PEDV) [14], and severe acute respiratory syndrome coronavirus 2 (SARS-CoV-2) [15]. Meanwhile, recombinase–aided amplification (RAA) has gained recognition as an advanced nucleic acid amplification technique. With its remarkable temperature adaptability, rapid reaction time, and simple operation, RAA is particularly suitable for diverse detection environments. Unlike conventional PCR methods, which require thermal cycling (denaturation, annealing, and extension), RAA achieves efficient amplification under isothermal conditions, making it highly valuable for agricultural pathogen detection [16,17].

In this study, we designed specific RAA primers and ssRNA reporter probes and developed a simple, specific, and sensitive novel detection system targeting the TGEV S gene by combining CRISPR/Cas13a technology with RAA. This integrated approach overcomes the limitations of traditional detection methods and is expected to provide valuable technical support for the early detection of TGEV.

2. Materials and Methods

2.1. Viruses and Samples

Porcine delta coronavirus (PDCoV), porcine reproductive and respiratory syndrome virus (PRRSV), porcine circovirus 2 type (PCV2), and porcine kobuvirus (PKoV) were previously isolated and stored in the National Center of Technology Innovation for Pigs. TGEV (strain SCJY–1), porcine epidemic diarrhea virus (PEDV) (strain SCSZ–1), and classical swine fever virus (CSFV) were isolated from vaccines produced by Chongqing Aolong Biological Products Co., Ltd. PRCV was kindly provided by the College of Animal Medicine, Southwest University. A total of 140 samples (including feces, intestinal tissues, and swabs) from porcine diarrhea cases were collected across different geographical locations, primarily from pig farms in the Luzhou, Dazhou, Longchang, Wulong, Changshou, Wanzhou, Banan, and Rongchang districts. All samples were stored at the National Animal Disease–Chongqing Monitoring Station and the Chongqing Academy of Animal Sciences.

2.2. Nucleic Acid Extraction

Viral RNA/DNA was extracted using the TaKaRa MiniBEST universal RNA/DNA extraction kit (TaKaRa, Dalian, China), following the manufacturer's instructions. The concentration of the extracted RNA was quantified using spectrophotometry, with A260/A280 ratios between 1.8 and 2.0 confirming RNA purity. The reverse–transcription reaction was performed using the PrimeScript™ RT Reagent Kit (TaKaRa, Dalian, China) to synthesize cDNA from the extracted RNA. The extracted nucleic acid and cDNA samples were stored at −80 °C.

2.3. Construction of Standard Recombinant Plasmid Containing Target Sequence

Since PRCV is a variant of TGEV, the whole genome sequences of TGEV and PRCV were downloaded from the GenBank database (http://www.ncbi.nlm.nih.gov/, accessed on 27 May 2024). Sequence alignment was carried out using the bioinformatic analysis software application MEGA 7.0, and DNA sequences present in the TGEV S gene but absent in PRCV were selected for subsequent studies. One of the consensus sequences with 100% alignment with the S gene was synthesized as the target sequence and cloned into a pUC57 vector (Sangon Biotech, Shanghai, China) to construct recombinant plasmids, named pUC57–TGEV. The plasmids were transformed into *E. coli* DH5α for cultivation. Positive plasmids were extracted using the E.Z.N.A Plasmid Mini–Extraction Kit (Omega Biotek Inc., Shanghai, China) and sequenced by Sangon Biotech Co., Ltd. The sequencing results confirmed the presence of the TGEV sequence. The recombinant plasmid was used as a standard positive control and stored at −20 °C for future use. The concentration of the extracted recombinant plasmid was determined using an ultraviolet spectrophotometer (OD260) (Thermo Fisher, Waltham, MA, USA), and the DNA copy number was computed as follows: dsDNA copy number (copies/μL) = (6.02 × 10^{23} (copies/mol) × concentration (ng/μL) × 10^{-9})/(DNA length × 660) [18,19].

2.4. Quantitative Real–Time PCR (qPCR) Assay

According to the requirements of the qPCR assay, one set of primers (F:5′-AGTCGTT AATGGATACCCA-3′, R: 5′-TGCACTCACTACCCCAATT-3′) and a probe (FAM-TCTGCT GAAGGTGCTATTAT-TAMRA) were designed from the conserved nucleotide region of the target sequence using Primer Premier 5.0 software (PREMIER Biosoft, San Francisco, CA, USA). The reaction system was 25 μL in volume, which included 12.5 μL of 2 × probe qPCR Mix with UNG (TaKaRa, Beijing, China), 0.5 μL of each F/R primer (10 μM), 1 μL of the probe, 2 μL of sample DNA/cDNA, and 8.5 μL of ddH_2O. All qPCR assays were carried out using a CFX96 Real-Time System (Bio-Rad, Shanghai, China) according to the following procedures: UNG action at 25 °C for 10 min, pre–denaturation at 95 °C for 30 s, 40 cycles of denaturation at 95 °C for 5 s, and annealing at 54 °C for 30 s, with the FAM fluorescent signal detected every 30 s. A series of 10–fold consecutive dilutions of the pUC57–TGEV plasmid were used as a template for the qPCR assay to construct the standard curves, which were generated based on the cycle threshold (Ct) values and the copy numbers (Log values) of the template DNA. The coefficients of determination (R^2) were calculated using the software application GraphPad Prism, v.8.0.1 (https://www.graphpad.com/scientific-software/prism/, accessed on 27 May 2024). The experiment was also performed three times independently to verify its repeatability. We evaluated the method's detection capability by obtaining 10–fold serial consecutive dilutions of the pUC57–TGEV plasmid as the template for the qPCR assay sensitivity test, and the detection limit of the latter was used as a template to conduct three duplicate tests with three duplicate wells, each for the determination of the repeatability of this method. Lastly, seven common porcine–derived nucleic acids of viruses, including TGEV, PRCV, PCV2, PKoV, PEDV, PRRSV, and

PDCoV, were selected as control templates for the assessment of the detection specificity of our method.

2.5. RAA Primers, ssRNA Reporter, and crRNA Preparation

Five RAA primers were designed online (https://ezassay.com/primer, accessed on 27 May 2024) based on the conserved nucleotide region of the TGEV S gene, according to the design requirements for RAA primers, which were synthesized by Sangon Biotech (Shanghai, China). A total of 2 forward primers and 3 reverse primers were paired, resulting in 6 pairs of primers, to determine the optimal primer combinations (Table 1). Basal RAA was performed with 10^4 copies/µL of the recombinant plasmid (pUC57–TGEV) as the template, according to the instructions provided with the RAA Nucleic Acid Amplification Kit (Zhuangbo, Guangxi, China). The best primer pair was identified by analyzing the brightness of the gel electrophoresis bands of 7 µL RAA products using the Image J 1.54 analysis tool. ddH_2O was used as a negative control.

Table 1. Information on recombinase–aided amplification (RAA) primers.

Number	Primer Sequences	Primer Pairing	Product Size, bp
F1	5′-GGTGTTAGGTGATTATTTTCCTACTGTACA-3′	F1/R1	308
F2	5′-ACGGTTAAACGTAGTCGTTAATGGATACCC-3′	F2/R1	164
F3	5′-CGGTTAAACGTAGTCGTTAATGGATACCCA-3′	F3/R1	163
R1	5′-ACCTGCACTCACTACCCCAATTGCAAGTCA-3′	F1/R2	309
R2	5′-AACCTGCACTCACTACCCCAATTGCAAGTC-3′	F2/R2	165
		F3/ R2	164

The crRNA (5′-GAUUUAGACUACCCCAAAAACGAAGGGGACUAAAACUC AGCAGAAU UAAAAUUGCGGGUUGUUG-3′) was designed and synthesized to target the RAA amplification product, according to the instructions for the enzyme LwaCas13a (Novoprotein, Suzhou, China) to ensure the proper functioning of the CRISPR/Cas13a reactive system; the 3′–end of the crRNA was ligated with palindromic repeat sequences (underlined) that bind to the Cas13a enzyme, forming a binary complex. Due to the cascade shearing activity of the Cas13a enzyme on RNA, the 5′- and 3′-ends of the ssRNA reporter were labeled with fluorescein FAM and quenching BHQ1 motifs, respectively, obtaining the sequence 5′-FAM–rUrUrUrUrUrU–BHQ1-3′. The RAA primers and crRNA were evaluated using a nucleotide BLAST + 2.15.0 search in the National Center for Biotechnology Information (NCBI) database, revealing no matches with sequences from other species.

2.6. CRISPR/Cas13a Complex Cleavage System

A cleavage premix system was prepared according to the CRISPR/Cas13a manual to verify the Cas13a cleavage activity and the accuracy of the crRNA. The 20.0 µL reaction system was composed of 2.0 µL of CRISPR/Cas13a 10 × buffer, 2.0 µL of LwaCas13a (200 pM), 8.0 µL of ddH_2O, 2.0 µL of crRNA (20 µM), 1.0 µL of the ssRNA reporter (20 mM), and 5.0 µL of the RNA template. The master mix was quickly placed into a real–time PCR instrument, where FAM channel fluorescence values were collected every 15 s at 37 °C for 40 min. Three replicates were set up for each cleavage reaction to obtain mean Ct values and identify optimal amplification, and the TGEV RNA template was replaced with ddH_2O as a negative control.

2.7. Establishment of TGEV Assay Based on RAA–CRISPR/Cas13a System

Based on the instructions of the fluorescein RAA nucleic acid amplification kit and the CRISPR /Cas13a manual, the RAA–CRISPR/Cas13a premix system for each reaction tube totaled 42.5 μL and consisted of 25.0 μL of A Buffer, 2.0 μL of the optimal forward primer with the T7 promoter (5′-GAAATTAATACGAC TCACTATAGGG-3′), 2.0 μL of the optimal reverse primer, 0.5 μL of LwaCas13a (5.0 pmol/μL), 0.25 μL of the crRNA (10.0 μM), 1.5 μL of T7 polymerase (50 U/μL; TaKaRa, Beijing, China), 2.0 μL of ribonucleotide triphosphate (rNTP) mix (50 mM, TaKaRa, Beijing, China), 2.0 μL of RNase inhibitor (40 U/μL; TaKaRa, Beijing, China), 0.5 μL of MgCl2 (500 mM), 0.5 μL of fluorescent ssRNA reporter (Bio–LifeSci, Shanghai, China), and 6.25 μL of RNase–free ddH_2O. After thorough mixing, 5.0 μL of standard plasmids at a concentration of 104 copies/μL was added to the reaction tube as a template; then, 2.5 μL of B Buffer was added to the inner cap of the assay tube, the cap was quickly closed, and the contents were mixed upside down and briefly centrifuged. The master mix was then placed into a real–time PCR instrument, where FAM channel fluorescence values were collected every 30 s at 37 °C for 40 min. Three replicates were set up for each cleavage reaction to obtain the mean Ct values and verify optimal amplification. ddH_2O was used as a negative control.

2.8. Optimization of TGEV Assay System Based on RAA–CRISPR/Cas13a

Given that RAA primer pairs, crRNA, and reporter RNA concentrations significantly influence the RAA–CRISPR/Cas13a reaction system, each element was classified into three levels, and the L9 (3^3) orthogonal test scheme was designed for the reaction system (Table 2). This optimized protocol uses the optimal primer pairs for fluorescence detection every 30 s for a total of 80 cycles. Each trial was repeated three times to obtain the mean Ct values.

Table 2. L9 (3^3) orthogonal analysis protocol design for the detection method based on RAA–CRISPR/Cas13a.

Test No.	Experimental Condition Parameters					
	A (Reporter RNA Concentration, μM)		B (Primer Concentration, μM)		C (crRNA Concentration, μM)	
1	1	8	1	8	1	2.5
2	1	8	2	10	3	10
3	1	8	3	12	2	5
4	2	10	1	8	3	10
5	2	10	2	10	2	5
6	2	10	3	12	1	2.5
7	3	12	1	8	2	5
8	3	12	2	10	1	2.5
9	3	12	3	12	3	10

2.9. Specificity Analysis Test of RAA–CRISPR/Cas13a Assay

Nucleic acids from TGEV, PCV2, PDCoV, PKoV, PEDV, PRRSV, and PRCV were used as templates to perform the specificity analysis of the TGEV assay based on RAA–CRISPR/Cas13a, assessing cross–reactivity with other pig–derived viruses or genomes. ddH_2O was used as a negative control.

2.10. Sensitivity Analysis Test of RAA–CRISPR/Cas13a Assay

The recombinant plasmid standard was consecutively diluted 10–fold with ddH_2O until a concentration of 10^{-2} copies/μL was reached. The sensitivity of the RAA–CRISPR/Cas13a assay was validated using the recombinant plasmid at 10^3–10^{-2} copies/μL as the standard nucleic acid template, with three replicates per reaction for each concentration to obtain the mean Ct values.

2.11. Stability Analysis Test of RAA–CRISPR/Cas13a Assay

Multiple replicates of the RAA–CRISPR/Cas13a assay were performed using the plasmid template at the detection limit for the evaluation of the stability of the detection method.

2.12. Effectiveness Evaluation of Testing Clinical Samples

A total of 140 clinical samples from diarrheic pigs were tested simultaneously to analyze the coincidence rate between the RAA–CRISPR/Cas13a assay, the RT–qPCR assay, and the RT–PCR assay for the further assessment of the reliability of the RAA–CRISPR/Cas13a assay in clinical detection.

2.13. Statistical Analysis

The statistical significance of the differences was determined using SPSSAU (https://spssau.com/index.html, accessed on 27 May 2024) to perform ANOVA and calculate the mean Ct values. Statistical analyses considered p-values < 0.05 as significant and p-values < 0.01 as highly statistically significant.

3. Results

3.1. Standard Plasmid Containing Target Sequence

The results of genome sequence alignment revealed significant differences in the S gene between TGEV and PRCV (Figure 1), consistent with previous reports [20,21]. A 362 bp consensus nucleic acid sequence (highlighted in the box), present in the S gene of TGEV but deleted in PRCV, was selected as the target for TGEV detection. This sequence was synthesized and ligated into the pUC57 vector, successfully constructing the recombinant plasmid pUC57–TGEV, whose sequencing results met expectations. The recombinant plasmid, extracted from positive clone bacteria, was measured to have a concentration of 139 ng/μL, corresponding to 4.13×10^{10} copies/μL.

3.2. Sensitivity and Specificity Validation of qPCR

A TaqMan qPCR method for detecting porcine TGEV was developed based on the screened target sequence of the S gene (sequence site: 20,517–20,879; GenBank ID: DQ811785). The detection limit of the qPCR method for the standard plasmid was 41.3 copies/μL (Table 3). The construction of the standard curves demonstrated a strong linear correlation ($R^2 = 0.9972$, $y = -3.4035\lg N + 39.429$) between the Ct values and the corresponding copy numbers of the target sequence in the TGEV S gene (Figure 2). Specificity evaluation revealed that the qPCR method produced an amplification curve exclusively for TGEV (Figure 3), and the coefficient of variation (CV) values for the nine detection tests at different concentrations of the standard plasmid were all below 1.5% (Table 3).

Table 3. Repeatability test results of qPCR.

Intra–Repetitive Assay			Inter–Repetitive Assay		
Average Value ($\bar{x}$)	Standard Deviation (s)	Coefficient of Variation (CV/%)	Average Value ($\bar{x}$)	Standard Deviation (s)	Coefficient of Variation (CV/%)
35.58	0.44	1.2	36.01	0.52	1.4
35.86	0.41	1.1			
36.60	0.42	1.1			

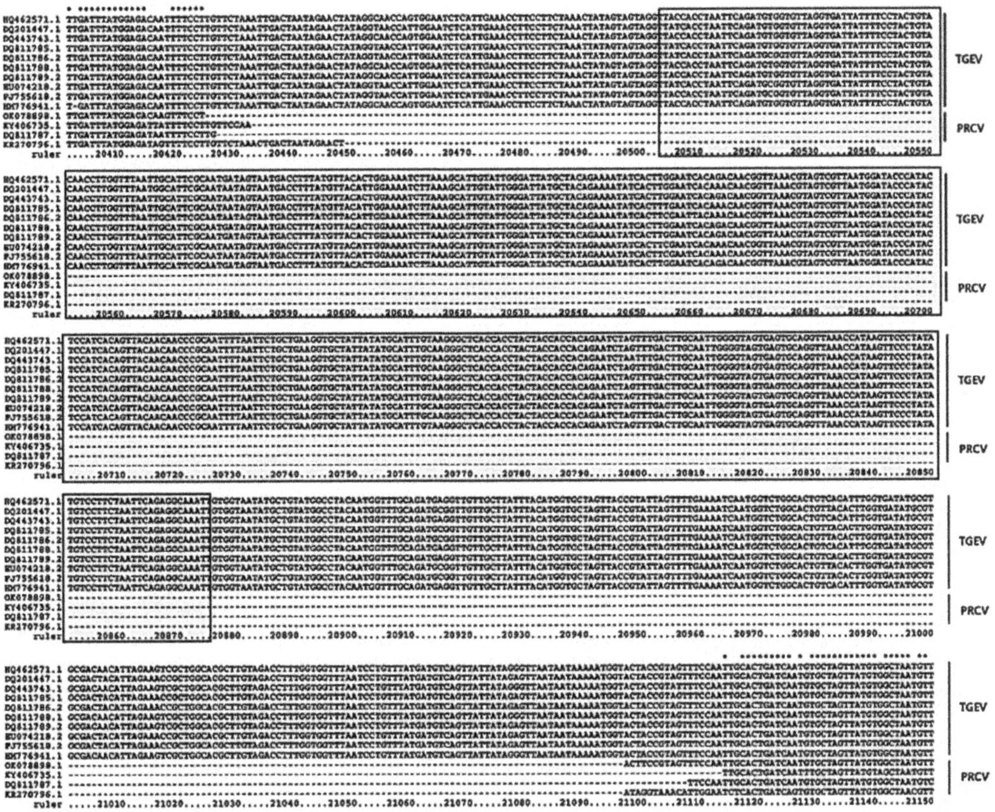

Figure 1. Partial sequence alignment of the S genes in different TGEV and PRCV strains; the consensus sequence within the box is the target sequence for detection. (Red mark) A 362 bp consensus nucleic acid sequence used for constructing the TGEV–puc57 recombinant plasmid.

3.3. Optimal Primer Screening for RAA

Six primer pairs were tested using basal RAA with 4.13×10^4 copies/μL of the pUC57–TGEV plasmid as the template to identify the optimal primer pairs for the method to ensure its specificity and effectiveness. The electrophoresis results showed that the sizes of the main bands of all RAA products were close to the theoretical values (Figure 4A), and the sequencing results met expectations. Only Lane 3 (F2/R1) and Lane 4 (F2/R2) displayed single bands with no non–specific bands. Further analysis of the gray values of the electrophoretic bands indicated that the band in Lane 3 was brighter than that in Lane 4 (Figure 4B). Based on these criteria, the optimal primer combination for RAA detection of TGEV is F2/R1.

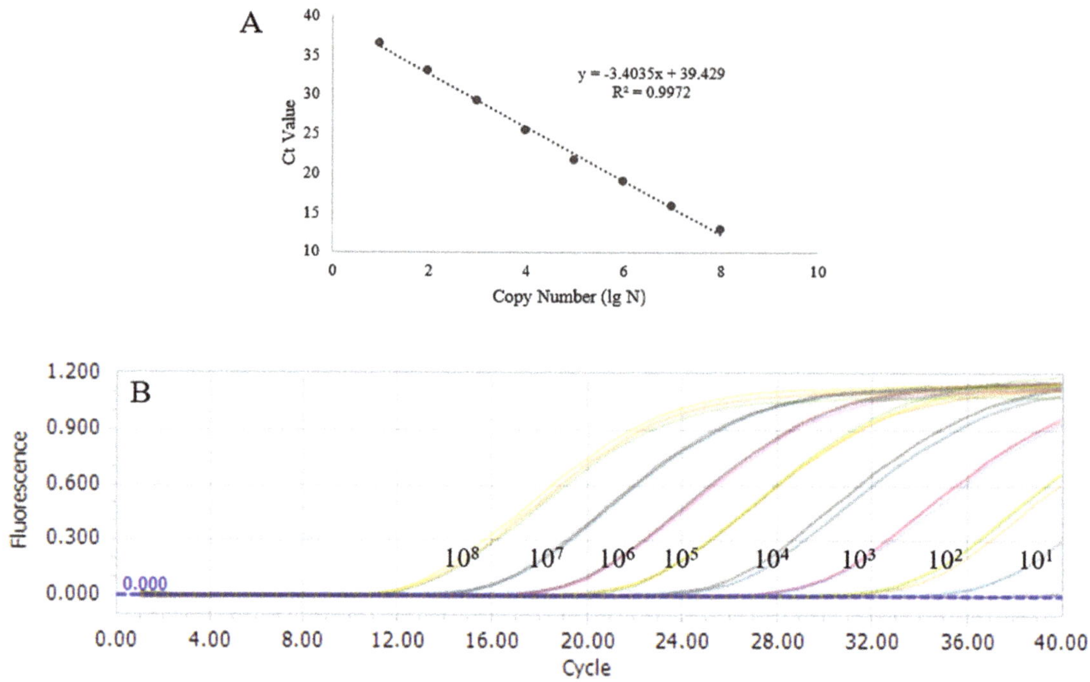

Figure 2. Sensitivity validation of qPCR. (**A**) Ten–fold serial dilutions of the TGEV plasmid were used as the template to construct the standard curve for qPCR. (**B**) Ten–fold serial dilutions of the pUC57–TGEV plasmid were used as the template for sensitivity detection in qPCR.

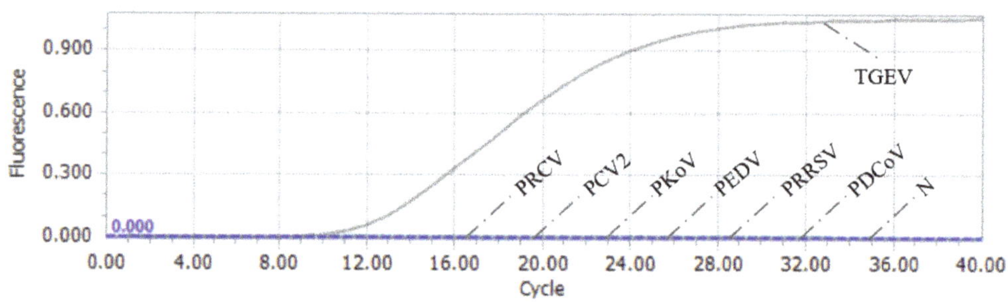

Figure 3. Specificity detection of qPCR.

3.4. *Detection of CRISPR/Cas13a Complex Cleavage*

For the CRISPR/Cas13a cleavage premix system, fluorescence signal values were successfully detected using a real–time PCR instrument (Figure 5). The results indicated that the template was TGEV RNA, and the fluorescence signal values gradually increased over time, validating the Cas13a protein cleavage activity and accuracy of the crRNA. This indicates that the Cas13a protein binds to the palindromic sequence attached to the crRNA, forming a binary complex. Once the crRNA correctly pairs with the detected target fragment of TGEV RNA, the Cas13a protein triggers its shearing activity, cleaving the six–base ssRNA reporters in the system and releasing fluorescent signals that can be captured.

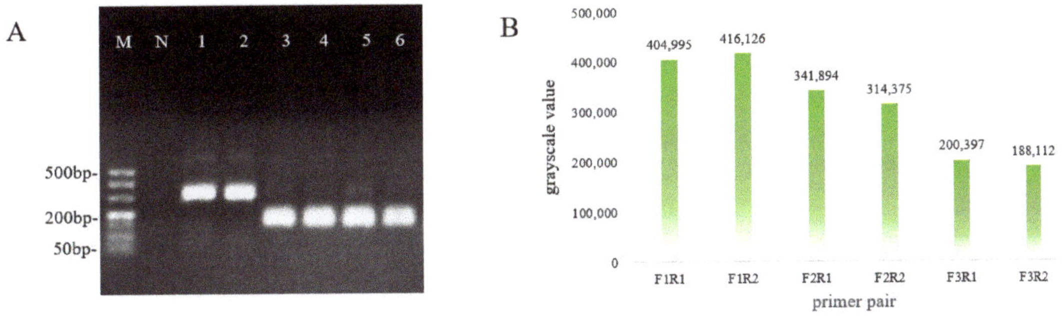

Figure 4. Determination of optimal primer pairs for the RAA method. (**A**) Optimal primer pairs were identified with the RAA method. (**B**) Image J analysis tool was used to evaluate the primer pairs and determine the most suitable one. M: DL500 DNA Marker; N: negative control; 1–6: primer pairs—F1/R1, F1/R2, F2/R1, F2/R2, F3/R1, and F3/R2, respectively.

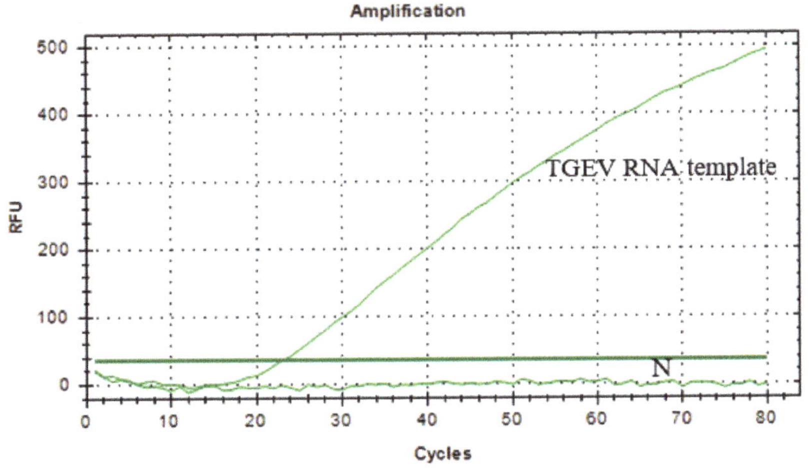

Figure 5. Detection results of the CRISPR/Cas13a complex cleavage system.

3.5. Orthogonal Test Optimization and Establishment of RAA–CRISPR/Cas13a Assay

The results of the L9 (3^3) orthogonal tests, along with their analysis, are presented in Tables 4 and 5. The k–values in Table 5 represent the average Ct values at different levels for each element, with a smaller k–value indicating higher detection efficiency and stability. Based on these reference conditions, the optimal combination is A3B2C3, suggesting that the optimal reaction conditions for efficiently amplifying TGEV nucleic acids are 12 μM ssRNA reporter, 10 μM crRNA, and 10 μM primer. The R values in Table 5 reflect the effect of each element on the RAA–CRISPR/Cas13a test, with larger R values indicating a greater influence of that element at its respective level. According to the R values from the L9 (3^3) results, the influence of the three elements is ordered as follows: A (ssRNA reporter) > C (crRNA) > B (RAA primer). The results of the three–factor analysis of variance (ANOVA) showed the value as $R^2 = 0.962$ (Table 6), indicating that the three factors—probe, primer, and crRNA—explain 99.6% of the variation in the Ct value, which means that the data from this experiment have statistical significance. Therefore, according to the above analysis, the optimized reaction system was established for the TGEV nucleic acid assay using the RAA–CRISPR/Cas13a method.

Table 4. Results of L9 (3^3) orthogonal tests.

Test No.	Experimental Condition Parameters			
	A (Probe)	B (Primer)	C (crRNA)	Ct Value
1	1	1	1	2.89
2	1	2	3	2.49
3	1	3	2	2.56
4	2	1	3	1.84
5	2	2	2	1.98
6	2	3	1	2.01
7	3	1	2	1.88
8	3	2	1	1.84
9	3	3	3	1.88

Table 5. Analysis of L9 (3^3) orthogonal test results.

	Results of Orthogonal Test		
	A (Probe)	B (Primer)	C (crRNA)
K1	7.93325	6.60910	6.74525
K2	5.83047	6.31189	6.41953
K3	5.60377	6.44651	6.20272
k1	2.64442	2.20303	2.24842
k2	1.94349	2.10396	2.13984
k3	1.86792	2.14884	2.06757
R	0.77649	0.09907	0.18084
Optimal conditions	A3	B2	C3

Table 6. Three–factor ANOVA results.

Source of Variation	*SS*	*df*	*MS*	*F*	*p*
Intercept	41.689	1	41.689	1801.233	0.001 **
A (ssRNA reporter concentration, μM)	1.109	2	0.554	23.957	0.040 *
B (primer concentration, μM)	0.015	2	0.008	0.325	0.755
C (crRNA concentration, μM)	0.047	2	0.024	1.026	0.494
Residual	0.046	2	0.023		

Note: $R^2 = 0.962$. * $p < 0.05$; ** $p < 0.01$.

3.6. Results of Specificity Assessment of RAA–CRISPR/Cas13a Assay

cDNA/DNA from TGEV, PCV2, PKoV, PEDV, PRRSV, PDCoV, and PRCV, all common pathogens that infect pigs, were used as nucleic acid templates for the assessment of the specificity of the RAA–CRISPR/Cas13a assay. As shown in Figure 6, the RAA–CRISPR/Cas13a assay did not produce a specific fluorescent signal for any pathogens other than TGEV, indicating that this method demonstrates high detection specificity for TGEV (Figure 6).

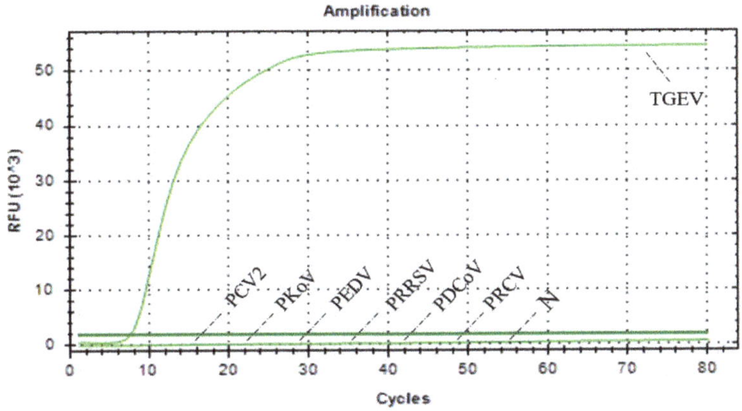

Figure 6. Results of the RAA–CRISPR/Cas13a specific test.

3.7. Results of Sensitivity Assessment of RAA–CRISPR/Cas13a Assay

A series of 10-fold dilutions of the pUC57–TGEV recombinant plasmid (ranging from 4.13 × 10^3 to 4.13 × 10^{-2} copies/μL) were tested as templates to investigate the sensitivity of the RAA–CRISPR/Cas13a assay. As shown in Figure 7, the results indicated that the minimum detectable copy number of pUC57–TGEV when using this method was 4.13 × 10^0 copies/μL (95% CI: 29.47–32.11 Ct). In comparison, the detection limit of the previously established qPCR method was 4.13 × 10^1 copies/μL, demonstrating that the RAA–CRISPR/Cas13a assay has higher detection sensitivity for TGEV.

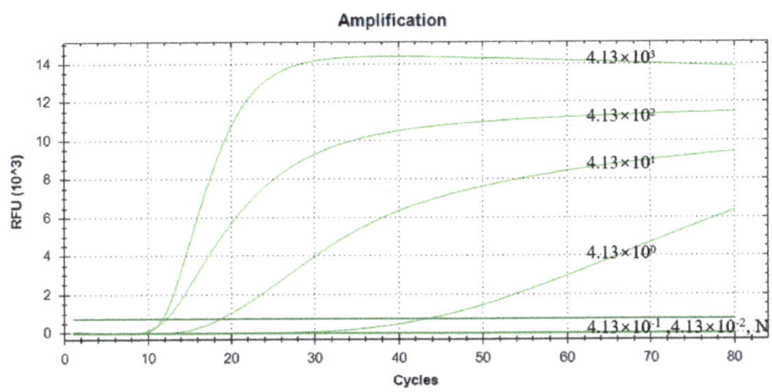

Figure 7. Sensitivity test results of plasmid templates with concentrations ranging from 4.13 × 10^3 to 4.13 × 10^{-2} copies/μL, detected with RAA–CRISPR/Cas13a.

3.8. Results of Stability Assessment of RAA–CRISPR/Cas13a Assay

For the plasmid templates with a detection limit of 4.13 × 10^0 copies/μL, results from three replicate experiments consistently showed similar, specific fluorescent signal curves and positive outcomes (Figure 8). This indicates that the RAA–CRISPR/Cas13a assay has good detection stability for TGEV.

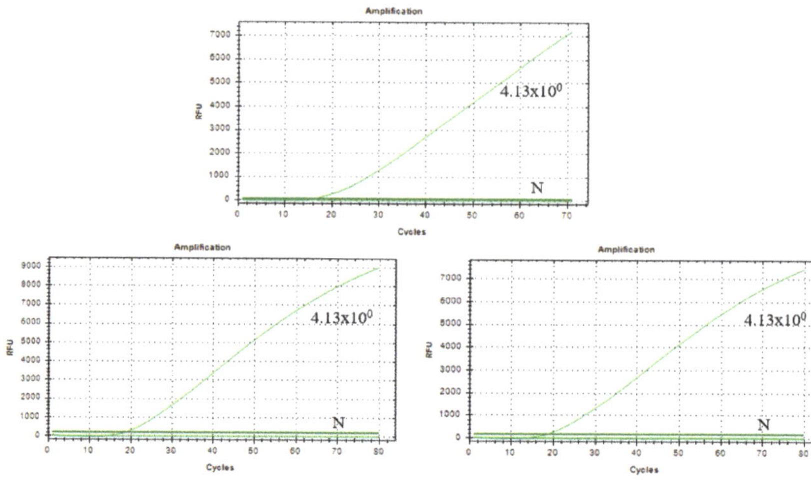

Figure 8. Results of the reproducibility test at a plasmid concentration of 4.13×10^0 copies/µL.

3.9. Clinical Sample Detection of RAA–CRISPR/Cas13a Assay

A total of 140 clinical samples were tested using both the RAA–CRISPR/Cas13a assay and the RT–qPCR and RT–PCR assays to evaluate the clinical application of the method. The comparative analysis of the test results is shown in Table 7. Both methods produced the same results for all clinical samples, indicating a 100% coincidence rate. Among them, a total of eight positive samples were detected with a positive rate of 5.7%, and the test results were completely consistent between the two detection methods.

Table 7. Comparison of test results for clinical samples obtained using the RAA–CRISPR/Cas13a assay, RT–qPCR method, and RT–PCR method.

No.	Region	Time	Quantity	Methods		
				RAA–CRISPR/Cas13a Detection Positive Results	RT–qPCR Detection Positive Results	RT–PCR Detection Positive Results
1	Wanzhou	19 February 2023	70	5/70	5/70	5/70
2	Luzhou	09 March 2023	4	1/4	1/4	1/4
3	Wulong	04 November 2023	20	2/20	2/20	2/20
4	Changshou	02 December 2023	8	0/8	0/8	0/8
5	Banan	17 December 2023	11	0/11	0/11	0/11
6	Rongchang	23 February 2024	16	0/16	0/16	0/16
7	Dazhou	21 November 2024	7	0/7	0/7	0/7
8	Longchang	24 December 2024	4	0/4	0/4	0/4

4. Discussion

TGE is a widely distributed, difficult–to–prevent, and difficult–to –control intestinal infectious disease that causes severe diarrhea in swine. Many countries and regions have reported outbreaks of TGE in swine herds, a condition that significantly impacts the gastrointestinal health of pigs and the development of the pig industry. Additionally, sick pigs are often co–infected and may harbor other diarrheal pathogens [22]. Notably, infected pigs with PRCV can experience mild TGEV onset at any age, which may obscure the presence of TGEV infection [6]. Early detection of TGEV is essential for implementing effective prevention and control measures. Reliable viral nucleic acid detection methods can offer accurate and efficient technical support for virus detection [23]. Therefore, in this study, we developed a novel TGEV nucleic acid assay based on RAA–CRISPR/Cas13a, focusing on the TGEV S gene.

Existing studies have shown that the CRISPR/Cas13a system can tolerate at most one mismatch between crRNA and target RNA [11,24]. Cas13a nuclease alone requires several hours to achieve amol sensitivity, necessitating a nucleic acid target pre–amplification step to increase the concentration for the CRISPR/Cas13a assay. RAA, which departs from traditional PCR techniques, can rapidly amplify nucleic acid targets in a short time. However, previous studies have indicated that RAA primers and probes can tolerate mismatches of up to nine bases with the target sequence [25], which significantly increases the risk of non–specific amplification due to these tolerable multi–base mismatches [26].

In view of the respective shortcomings of the two methods described above, we established the RAA–CRISPR/Cas13a assay. We began by designing and screening specific RAA primers and corresponding crRNA based on the conserved sequence of the TGEV S gene to avoid interference from PRCV nucleic acid. Given the limited number of Cas13a proteins and the potential for abundant crRNA to compete with the ssRNA reporter for cleavage [27–29], we further optimized the RAA–CRISPR/Cas13a assay through orthogonal experiments. This optimization focused on three factors: primer concentration, which can affect the amplification efficiency of RAA, and crRNA concentration and ssRNA reporter concentration, both of which can affect the cleavage efficiency of the Cas13a protease in the reaction system.

The optimized RAA–CRISPR/Cas13a assay represents a significant advancement over LAMP methods for TGEV detection. Although LAMP has been widely adopted for its field adaptability and simple equipment requirements, the CRISPR–based system demonstrates superior performance, with higher sensitivity (4.13 copies/μL vs. LAMP's typical 10–100 copies/μL), specificity through RNA–guided target recognition, and a faster turnaround time (40 min). The closed–tube design of the RAA–CRISPR/Cas13a assay additionally minimizes contamination risks that can affect open–tube LAMP reactions [30–32]. Current CRISPR implementation faces challenges in field deployment due to fluorescence detection requirements and higher reagent costs, while its exceptional detection capabilities make it particularly valuable for laboratory–based confirmatory testing and settings where maximum accuracy is paramount. The field application of the RAA–CRISPR/Cas13a assay is expected to be realized using portable fluorescence detectors or lateral flow chromatographic test strips, offering more efficient and practical solutions for the clinical detection of TGEV [33,34].

Author Contributions: Methodology and writing—original draft, L.L.; validation, H.M.; formal analysis, S.L.; software, J.G.; data curation, M.L. and S.N.; conceptualization, G.X.; resources, L.F.; writing—review and editing, Z.S. and L.Y. All authors have read and agreed to the published version of the manuscript.

Funding: This research study was supported by the National Center of Technology Innovation for Pigs (NCTIP–XD/B11 and NCTIP–XD/B19), Chongqing Talent plan "contract system" project (cstc2022ycjh–bgzxm0183). The funders had no role in the design of the study; in the collection, analyses, or interpretation of data; in the writing of the manuscript; or in the decision to publish the results.

Institutional Review Board Statement: Not applicable.

Informed Consent Statement: Not applicable.

Data Availability Statement: The raw datasets used and/or analyzed in the current study are available from the corresponding authors upon request.

Conflicts of Interest: The authors declare no conflicts of interest.

Abbreviations

The following abbreviations are used in this manuscript:

ANOVA	Analysis of variance
ASFV	African swine fever virus
CRISPR	Clustered regularly interspaced short palindromic repeat
Cas	CRISPR–associated proteins
CV	Coefficient of variation
CSFV	Classical swine fever virus
PCV2	Porcine circovirus 2
PDCoV	Porcine delta coronavirus
PEDV	Porcine epidemic diarrhea virus
PKoV	Porcine kobuvirus
PRCV	Porcine respiratory coronavirus
PRRSV	Porcine reproductive and respiratory syndrome virus
SARS-CoV-2	Severe acute respiratory syndrome coronavirus 2
TGEV	Porcine transmissible gastroenteritis virus
RAA	Recombinase–aided amplification

References

1. Kim, L.; Chang, K.O.; Sestak, K.; Parwani, A.; Saif, L.J. Development of a reverse transcription–nested polymerase chain reaction assay for differential diagnosis of transmissible gastroenteritis virus and porcine respiratory coronavirus from feces and nasal swabs of infected pigs. *J. Vet. Diagn. Investig.* **2000**, *12*, 385–388. [CrossRef] [PubMed]
2. Casanova, L.; Rutala, W.A.; Weber, D.J.; Sobsey, M.D. Survival of surrogate coronaviruses in water. *Water Res.* **2009**, *43*, 1893–1898. [CrossRef] [PubMed]
3. Garwes, D.J. Transmissible gastroenteritis. *Vet. Rec.* **1988**, *122*, 462–463. [CrossRef] [PubMed]
4. Park, S.; Sestak, K.; Hodgins, D.C.; Shoup, D.I.; Ward, L.A.; Jackwood, D.J.; Saif, L.J. Immune response of sows vaccinated with attenuated transmissible gastroenteritis virus (Tgev) and recombinant Tgev spike protein vaccines and protection of their suckling pigs against virulent Tgev challenge exposure. *Am. J. Vet. Res.* **1998**, *59*, 1002–1008. [CrossRef]
5. Chen, Y.; Zhang, Y.; Wang, X.; Zhou, J.; Ma, L.; Li, J.; Yang, L.; Ouyang, H.; Yuan, H.; Pang, D. Transmissible Gastroenteritis Virus: An Update Review and Perspective. *Viruses* **2023**, *15*, 359. [CrossRef]
6. Magtoto, R.; Poonsuk, K.; Baum, D.; Zhang, J.; Chen, Q.; Ji, J.; Piñeyro, P.; Zimmerman, J.; Giménez–Lirola, L.G. Evaluation of the Serologic Cross–Reactivity between Transmissible Gastroenteritis Coronavirus and Porcine Respiratory Coronavirus Using Commercial Blocking Enzyme–Linked Immunosorbent Assay Kits. *mSphere* **2019**, *4*, e00017-19. [CrossRef]
7. Woods, R.D.; Wesley, R.D.; Kapke, P.A. Neutralization of porcine transmissible gastroenteritis virus by complement–dependent monoclonal antibodies. *Am. J. Vet. Res.* **1988**, *49*, 300–304. [CrossRef] [PubMed]
8. Saif, L.J. Comparative pathogenesis of enteric viral infections of swine. *Adv. Exp. Med. Biol.* **1999**, *473*, 47–59. [CrossRef]
9. Saif, L.J.; Jung, K. Comparative Pathogenesis of Bovine and Porcine Respiratory Coronaviruses in the Animal Host Species and SARS-CoV-2 in Humans. *J. Clin. Microbiol.* **2020**, *58*, e01355-20. [CrossRef]
10. Gootenberg, J.S.; Abudayyeh, O.O.; Lee, J.W.; Essletzbichler, P.; Dy, A.J.; Joung, J.; Verdine, V.; Donghia, N.; Daringer, N.M.; Freije, C.A.; et al. Nucleic acid detection with Crispr–Cas13a/C2c2. *Science* **2017**, *356*, 438–442. [CrossRef]
11. Abudayyeh, O.O.; Gootenberg, J.S.; Konermann, S.; Joung, J.; Slaymaker, I.M.; Cox, D.B.; Shmakov, S.; Makarova, K.S.; Semenova, E.; Minakhin, L.; et al. C2c2 is a single–component programmable RNA–guided RNA–targeting CRISPR effector. *Science* **2016**, *353*, aaf5573. [CrossRef] [PubMed]
12. Liu, L.; Li, X.; Ma, J.; Li, Z.; You, L.; Wang, J.; Wang, M.; Zhang, X.; Wang, Y. The Molecular Architecture for RNA–Guided RNA Cleavage by Cas13a. *Cell* **2017**, *170*, 714–726.e10. [CrossRef] [PubMed]
13. Wang, Z.; Wang, Y.; Zhang, Y.; Qin, G.; Sun, W.; Wang, A.; Wang, Y.; Zhang, G.; Zhao, J. On–site detection and differentiation of African swine fever virus variants using an orthogonal CRISPR–Cas12b/Cas13a–based assay. *iScience* **2024**, *27*, 109050. [CrossRef] [PubMed]
14. Liu, H.; Yin, D.D.; Shao, Y.; Song, X.; Wang, Z.; Pan, X.; Tu, J.; He, C.; Zhu, L.; Qi, K.; et al. Establishment and preliminary application of RAA–CRISPR/Cas13a assay for porcine epidemic diarrhea virus. *J. Anim. Husb. Vet. Sci.* **2023**, *54*, 3991–3997.
15. Wang, L.; Zhou, J.; Wang, Q.; Wang, Y.; Kang, C. Rapid design and development of CRISPR–Cas13a targeting SARS-CoV-2 spike protein. *Theranostics* **2021**, *11*, 649–664. [CrossRef] [PubMed] [PubMed Central]

16. Wang, Z.H.; Li, P.; Lin, X.; Jia, H.; Jiang, Y.T.; Wang, X.J.; Hou, S.H. Application of portable real–time recombinase–aided amplification (rt–RAA) assay in the clinical diagnosis of ASFV and prospective DIVA diagnosis. *Appl. Microbiol. Biotechnol.* **2021**, *105*, 3249–3264. [CrossRef]
17. Wang, Q.Y.; Li, F.; Shen, X.X.; Fu, S.H.; He, Y.; Lei, W.W.; Liang, G.D.; Wang, H.Y.; Ma, X.J. A Reverse–transcription Recombinase–aided Amplification Assay for the Rapid Detection of the Far–Eastern Subtype of Tick–borne Encephalitis Virus. *Biomed. Environ. Sci.* **2019**, *32*, 357–362. [CrossRef]
18. Lu, Q.; Zhao, Z.Y.; Yin, D.W.; Liu, Y.; Wang, W.; Zheng, M.; Hu, S.; Zhao, C.; Zhang, X.; Lei, X.; et al. Establishment and preliminary application of fluorescent RT–RAA method for porcine infectious gastroenteritis virus. *J. Anim. Sci. Vet. Med.* **2023**, *54*, 2208–2214.
19. Lu, L.D.; Mou, H.; Hu, X.; Liu, M.; Li, S.; Li, X.; Song, Z.; Yang, L. Establishment and preliminary application of RAA detection method for S gene of porcine transmissible gastroenteritis virus. *Chin. J. Anim. Husb. Vet. Med.* **2024**, *55*, 3590–3599.
20. Turlewicz, P.H.; Pomorska, M.M. Porcine Coronaviruses: Overview of the State of the Art. *Virol. Sin.* **2021**, *36*, 833–851. [CrossRef]
21. Chen, F.; Knutson, T.P.; Rossow, S.; Saif, L.J.; Marthaler, D.G. Decline of transmissible gastroenteritis virus and its complex evolutionary relationship with porcine respiratory coronavirus in the United States. *Sci. Rep.* **2019**, *9*, 3953. [CrossRef] [PubMed]
22. Zhang, Q.; Hu, R.; Tang, X.; Wu, C.; He, Q.; Zhao, Z.; Chen, H.; Wu, B. Occurrence and investigation of enteric viral infections in pigs with diarrhea in China. *Arch. Virol.* **2013**, *158*, 1631–1636. [CrossRef] [PubMed]
23. Zhang, M.L.; Huang, F.Q.; Zuo, X.L.; Liang, J.; Liang, K.; Shan, J.; Li, Z.; Yu, J.; Luo, L.; Xuan, Z.; et al. Establishment of an elasmobranch virus detection method for largemouth bass based on CRISPR/Cas13a system. *J. Aquat.* **2024**, *48*, 283–291. [CrossRef]
24. Zhang, Y.X.; Wang, S.M.; Duan, H.Y.; Zhao, K.; Dong, S.; Gao, F.; Li, L. Establishment and application of TaqMan real–time PCR for detection of porcine infectious gastroenteritis virus. *China Vet. Sci.* **2024**, *4*, 479–484. [CrossRef]
25. Boyle, D.S.; Lehman, D.A.; Lillis, L.; Peterson, D.; Singhal, M.; Armes, N.; Parker, M.; Piepenburg, O.; Overbaugh, J. Rapid detection of HIV–1 proviral DNA for early infant diagnosis using recombinase polymerase amplification. *MBio* **2013**, *4*, e00135-13. [CrossRef]
26. Yu, J.X.; Wei, S.S.; Qin, S.M.; Wu, J.; Yang, L.; Chen, F.; Xu, L.; Qin, S.; Hua, J.; Wei, J.; et al. Establishment of a Rapid Detection Method for Porcine Reproductive and Respiratory Syndrome Virus Based on CRISPR/Cas12a–RT–RAA. *China Anim. Husb. Vet. Med.* **2024**, *51*, 3237–3246. [CrossRef]
27. Huang, Z.; Tian, D.; Liu, Y.; Lin, Z.; Lyon, C.J.; Lai, W.; Fusco, D.; Drouin, A.; Yin, X.; Hu, T.; et al. Ultra–sensitive and high–throughput CRISPR–p owered COVID–19 diagnosis. *Biosens. Bioelectron.* **2020**, *164*, 112316. [CrossRef]
28. Yin, L.; Man, S.; Ye, S.; Liu, G.; Ma, L. CRISPR–Cas based virus detection: Recent advances and perspectives. *Biosens. Bioelectron.* **2021**, *193*, 113541. [CrossRef]
29. Li, L.; Li, S.; Wu, N.; Wu, J.; Wang, G.; Zhao, G.; Wang, J. HOLMESv2: A CRISPR–Cas12b–assisted platform for nucleic acid detection and DNA methylation quantitation. *ACS Synth. Biol.* **2019**, *8*, 2228–2237. [CrossRef]
30. Kellner, M.J.; Koob, J.G.; Gootenberg, J.S.; Abudayyeh, O.O.; Zhang, F. SHERLOCK: Nucleic acid detection with CRISPR nucleases. *Nat. Protoc.* **2019**, *14*, 2986–3012. [CrossRef]
31. Das, D.; Lin, C. –W.; Chuang, H.–S. LAMP–Based Point–of–Care Biosensors for Rapid Pathogen Detection. *Biosensors* **2022**, *12*, 1068. [CrossRef] [PubMed]
32. Ma, B.; Li, J.; Chen, K.; Yu, X.; Sun, C.; Zhang, M. Multiplex Recombinase Polymerase Amplification Assay for the Simultaneous Detection of Three Foodborne Pathogens in Seafood. *Foods* **2020**, *9*, 278. [CrossRef] [PubMed]
33. Zhu, T.; Jiang, W.; Wu, Y.; Fang, R.; Deng, F.; Yang, D. Advances in CRISPR/Cas13a–based biosensors for non–coding RNA detection. *Talanta* **2025**, *294*, 128223. [CrossRef] [PubMed]
34. Yin, J.; Cui, J.; Zheng, H.; Guo, T.; Wei, R.; Sha, Z.; Gu, S.; Ni, B. Implementation of RT–RAA and CRISPR/Cas13a for an NiV Point–of–Care Test: A Promising Tool for Disease Control. *Viruses* **2025**, *17*, 483. [CrossRef]

veterinary sciences

MDPI

Review

A Review of Cross-Species Transmission Mechanisms of Influenza Viruses

Xianfeng Hui [1,2,3,†], Xiaowei Tian [4,†], Shihuan Ding [1,2,3], Ge Gao [2], Jiyan Cui [1,2,3], Chengguang Zhang [5], Tiesuo Zhao [1,3], Liangwei Duan [2,6] and Hui Wang [2,6,*]

1 Department of Immunology, School of Basic Medical Sciences, Xinxiang Medical University, Xinxiang 453003, China; xianfenghui@163.com (X.H.)
2 Henan Key Laboratory of Immunology and Targeted Drug, Xinxiang Medical University, Xinxiang 453003, China
3 Xinxiang Engineering Technology Research Center of Immune Checkpoint Drug for Liver-Intestinal Tumors, Xinxiang Medical University, Xinxiang 453003, China
4 Department of Pathogenic Biology, School of Basic Medical Sciences, Xinxiang Medical University, Xinxiang 453003, China
5 National Key Laboratory of Agricultural Microbiology, Huazhong Agricultural University, Wuhan 430070, China
6 Henan Collaborative Innovation Center of Molecular Diagnosis and Laboratory Medicine, School of Medical Technology, Xinxiang Medical University, Xinxiang 453003, China
* Correspondence: wanghui@xxmu.edu.cn
† These authors contributed equally to this work.

Simple Summary: The cross-species transmission of influenza viruses is pivotal in zoonotic pandemics, driven by viral genetic adaptability, host–receptor compatibility, and ecological factors. Key mutations in hemagglutinin and neuraminidase enable host barrier breaches, while intermediate hosts (e.g., pigs) act as viral "mixers" for strain diversification. Advances in sequencing and structural biology have identified critical molecular markers, yet challenges persist in deciphering dynamic virus–host evolution, establishing real-time surveillance, and designing broad-spectrum interventions. Integrating multidisciplinary strategies—including One Health networks and artificial intelligence-driven prediction models—is crucial for building a multi-layered defense system. This review synthesizes current progress and challenges, offering a framework to optimize pandemic preparedness against influenza spillover risks.

Abstract: The cross-species transmission of influenza viruses represents a critical link in the pandemic of zoonotic diseases. This mechanism involves multi-level interactions, including viral genetic adaptability, host–receptor compatibility, and ecological drivers. Recent studies have highlighted the essential role of mutations in hemagglutinin and neuraminidase in overcoming host barriers, while elucidating the differences in the distribution of host sialic acid receptors. Furthermore, the "mixer" function of intermediate hosts, such as pigs, plays a significant role in viral redistribution. Advances in high-throughput sequencing and structural biology technologies have gradually resolved key molecular markers and host restriction factors associated with these viruses. However, challenges remain in understanding the dynamic evolutionary patterns of virus–host interaction networks, developing real-time early warning capabilities for cross-species transmission, and formulating broad-spectrum prevention and control strategies. Moving forward, it is essential to integrate multidisciplinary approaches to establish a multi-level defense system, leveraging the 'One Health' monitoring network, artificial intelligence prediction models, and new vaccine research and development to address the ongoing threat of cross-species transmission of influenza viruses. This paper systematically reviews the research progress and discusses

Vet. Sci. **2025**, *12*, 447 https://doi.org/10.3390/vetsci12050447

bottlenecks in this field, providing a theoretical foundation for optimizing future prevention and control strategies.

Keywords: influenza virus; cross-species transmission; mechanisms; One Health

1. Introduction

Influenza viruses pose a significant threat to both human and animal health, characterized by their ability to cross species barriers and cause epidemics and pandemics [1,2]. The global status of animal influenza has been documented through extensive surveillance, revealing a rich diversity of influenza subtypes across various host species [3]. A comprehensive database compiled up to 2016 identified over 70,000 records of animal influenza events, highlighting the ongoing risk of new subtypes emerging from animal reservoirs, particularly in regions like Asia, North America, and Europe, where the diversity of influenza viruses is greatest [4]. This diversity is a critical factor in the potential for reassortment, which can lead to the emergence of novel strains with pandemic potential [4].

In recent years, significant progress has been made in understanding the mechanisms of cross-species transmission of influenza viruses. Studies have demonstrated that avian influenza viruses (AIVs) possess the ability to adapt to mammalian hosts through specific mutations that enhance their capacity to bind to human-type receptors [5,6]. For instance, the H7N9 avian influenza virus has been identified as dual-receptor tropic, indicating a limited potential for human-to-human transmission, although certain mutations may augment this capability [7–9]. Similarly, by analyzing the changes in the receptor-binding specificity of influenza viruses to host receptors, researchers can trace the adaptive evolution of swine influenza viruses, as this specificity is critical for their efficient transmission among pig populations. The mechanisms underlying the interspecies transmission of influenza viruses are complex and involve various factors, including the host's immune response and the virus's ability to evade these defenses [10]. Type I interferons (IFNs) play a crucial role in the host's defense against influenza virus infections, but influenza viruses have evolved strategies to counteract these responses [11,12], complicating the dynamics of transmission. Additionally, the polymerase basic protein 2 (PB2) has been implicated in the adaptation of influenza viruses to new hosts, with specific mutations influencing the virus's ability to replicate and transmit effectively [13,14].

Among the diverse viral families with cross-species transmission potential, influenza viruses are particularly concerning due to their ability to jump between hosts and generate novel, potentially pandemic strains. While influenza B and C viruses exhibit limited zoonotic potential due to their restricted host ranges—primarily affecting humans and seals—and relatively stable genomic architectures [15,16], influenza A viruses (IAVs) demonstrate unparalleled cross-species versatility. Genomic surveillance has identified IAV infections across 12 mammalian orders, ranging from cetaceans to primates [17], as well as in all major avian taxa [18]. This remarkable host breadth is driven by two primary evolutionary features: first, the virus's co-evolution with aquatic wild birds, which serve as an ancient and genetically diverse reservoir [18,19]; and second, the capacity for adaptive mutations in the viral hemagglutinin (HA) protein, which enable flexible receptor binding and facilitate interspecies transmission [20]. These properties position IAV as a central model for studying zoonotic emergence and pandemic risk. The pandemic potential of IAVs is also influenced by their genetic diversity and the ability to undergo rapid evolution [21]. Continuous surveillance and research are essential to identify viruses with pandemic potential before they fully adapt to human populations. This includes understanding the ecological factors that facilitate the emergence of new strains, such as the close contact between humans and animals in agricultural settings [22]. Moreover, the role of exposed swine in the transmission of influenza A viruses to humans has been highlighted, with cases often occurring among individuals with close contact to swine [23]. This underscores the importance of biosecurity measures in preventing the spread of influenza viruses across species [24]. In summary, the epidemic status of influenza viruses is marked by their ongoing evolution and potential for cross-species transmission, necessitating robust surveillance and research efforts to mitigate the risks associated with these viruses [25]. Understanding the mechanisms of transmission and adaptation is crucial for developing effective prevention strategies and vaccines to combat future influenza pandemics. We aim to summarize the current knowledge about influenza, including the classification, epidemic status, cross-species dissemination mechanisms, public health importance, and prospects and challenges of cross-species transmission of influenza viruses.

2. Classification of Influenza Viruses

The influenza virus is an enveloped negative-strand RNA virus that belongs to the genus Influenza within the Orthomyxoviridae family. Based on differences in the nucleoprotein (NP) and matrix protein (M), the virus can be categorized into four types: A, B, C, and D [26–28].

Among these pathogens, the influenza A virus stands out as the most common threat to both humans and animals, leading to what is commonly known as seasonal flu [16]. Additionally, the influenza A virus was responsible for four significant pandemics: the 1918 Spanish influenza, the 1957 Asian influenza, the 1968 Hong Kong influenza, and the 2009 swine-origin pandemic influenza (2009 pH1N1). It was also the cause of a comparatively milder pandemic, known as the 1977 Russian influenza [29–31]. Regarding the origins, development, and ecological aspects of influenza viruses, migratory birds, including shorebirds and waterfowl, are considered major reservoir hosts, offering a diverse array of viral gene segments that can lead to the emergence of new reassortant viruses [18,19,32]. Concurrently, pigs are widely regarded as optimal "mixing vessels" for influenza A viruses due to their susceptibility to both avian- and human-origin strains. This is facilitated by the co-expression of α2,3- and α2,6-linked sialic acid receptors in the porcine respiratory epithelium [33], allowing for co-infection and genetic reassortment. A prominent example is the 2009 H1N1 pandemic virus, which arose through triple reassortment involving avian, human, and swine influenza lineages (Figure 1) [34].

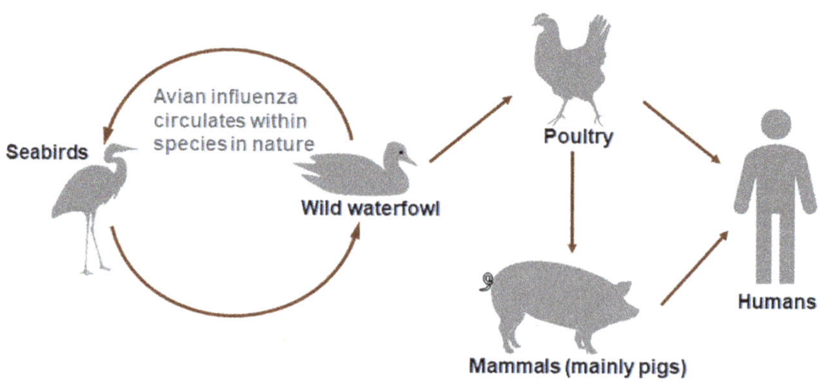

Figure 1. Transmission routes of influenza epidemics.

Influenza viruses that circulate among human populations include the influenza B virus (IBV) in addition to IAV [35]. Both of these virus types are included in the formulations of current seasonal influenza vaccines. Historically, IBV has been responsible for approximately 20% of influenza-related hospitalizations each year and can dominate specific influenza seasons, as observed in Europe during 2017/2018 [35]. The severity of illness caused by IBV is similar to that caused by IAV, particularly affecting children and young adults who are most vulnerable to IBV [36,37]. In contrast to IAV, which is typically found in aquatic bird populations, IBV infections are primarily confined to humans, with very few instances of infection reported in seals [15]. However, mice and ferrets can be experimentally infected with IBV, making these animals valuable models for studying human IBV infection and related diseases [38,39]. Due to its restricted host range and a significantly slower mutation rate compared to IAV, circulating IBV strains show less antigenic diversity when compared to H1N1 and H3N2 IAV strains [40,41]. Despite this, since the initial reports in the 1940s, IBV has gradually evolved into two separate lineages—B/Victoria/2/87-like and B/Yamagata/16/88-like—henceforth referred to as the B/Victoria and B/Yamagata lineages, respectively, which are further subdivided into distinct antigenic clades [36,42].

Unlike influenza A and B viruses that infect humans and cause severe diseases in seasonal epidemics, the influenza C virus (ICV) is a ubiquitous childhood pathogen typically causing mild respiratory symptoms [43,44]. Additionally, ICVs can infect various animals, including pigs, dogs, and cattle [45]. However, ICVs are less likely to present a significant threat to human health [46,47]. Influenza D virus (IDV) is considerably rarer, with no documented cases of human infection. IDVs primarily infect a limited range of animals, including pigs and cattle, and have a minimal impact on public health [48]. Currently, there are no recognized subtypes of influenza C and D viruses.

3. Epidemic Status of Influenza A Virus

Influenza viruses pose significant threats to both agricultural sectors and human health, primarily due to their ability to infect a wide range of hosts, including birds, pigs, and humans [21]. The epidemiology of these viruses is complex, characterized by frequent interspecies transmission and reassortment, which can lead to the emergence of novel strains with pandemic potential [49,50].

It is estimated that seasonal infections caused by the influenza virus lead to approximately 250,000 to 500,000 deaths each year globally [51]. In healthy adults, seasonal influenza infections are typically self-limiting and mainly confined to the upper respiratory system; however, these infections can be severe in children and the elderly, potentially resulting in viral pneumonia. Besides humans, the A strains of the influenza virus have the capability to infect a variety of host species, including waterfowl, pigs, domesticated birds, and seals. Consequently, influenza A viruses present in zoonotic reservoirs have sporadically led to widespread infection and even pandemics among humans [21]. The most recent four pandemics related to influenza—namely, the 1918 H1N1 Spanish flu [52,53], the 1957 H2N2 Asian flu [54,55], the 1968 H3N2 Hong Kong flu [56], and the 2009 H1N1 [34,57]—were due to the transmission of influenza A virus from zoonotic reservoirs to humans (Figure 2). Additionally, strains of the influenza A virus, such as H5N1, H7N7, and H7N9, have successfully crossed the species barrier from domestic poultry to induce lethal infections in humans [58–60].

In agriculture, particularly in the poultry and swine industries, IAV has been responsible for severe economic losses [61,62]. Avian influenza viruses can be divided into highly pathogenic avian influenza (HPAI) viruses and low pathogenic avian influenza (LPAI) viruses [63]. Highly pathogenic avian influenza viruses are limited to the H5 and

H7 subtypes, which cause severe illness and high mortality rates in gallinaceous species. In contrast, low pathogenic avian influenza (LPAI) viruses are typically maintained in wild aquatic birds and can be transmitted to domestic poultry species, often resulting in subclinical infections or mild respiratory diseases [64].

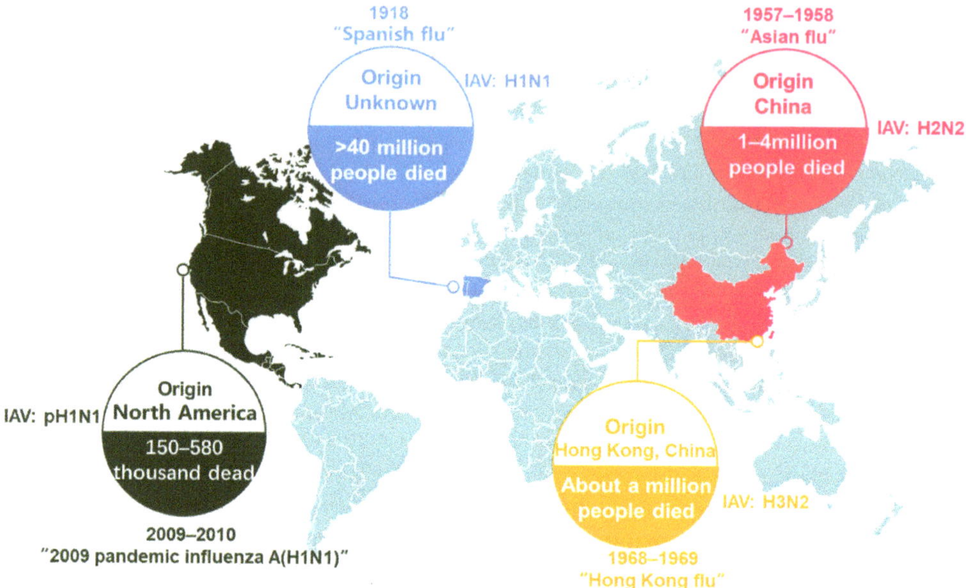

Figure 2. Four global flu pandemics recorded since the early twentieth century.

The H5N1, a highly pathogenic avian influenza (HPAI) subtype of influenza A virus, was initially identified in domestic geese in Guangdong, China, in 1996 and has since been responsible for numerous outbreaks among bird populations [65]. Starting in 2020, a variant from the evolutionary lineage 2.3.4.4b of the H5 avian influenza virus has devastated many regions across Africa, Asia, and Europe, resulting in unprecedented mortality rates among wild birds and poultry. The virus reached North America in 2021 and spread to Central and South America in 2022 [66]. In 2022, 67 nations across five continents reported occurrences of H5N1 highly pathogenic avian influenza (HPAI) in both poultry and wild bird populations to the World Organization for Animal Health. The death or culling of over 131 million poultry was recorded on impacted farms and in communities. By 2023, as the virus further propagated, outbreaks were noted in an additional 14 countries, primarily in the Americas. In several instances of reported mass fatalities among wild birds, the causative agent has been confirmed as the influenza A H5N1 clade 2.3.4.4b virus [67].

Avian influenza viruses are widely recognized for their transmission among birds; however, there has been a notable increase in the detection of H5N1 avian influenza viruses in mammals. In April 2024, a dairy farm worker in Texas was diagnosed with H5N1 virus infection after contact with infected cows. This case represents the first documented instance of cattle transmission of the virus, with the primary symptom reported by the patient being conjunctivitis. As of 17 January 2025, the United States has confirmed 67 cases of highly pathogenic H5N1 virus infection, the majority of which are associated with contact with cattle [68]. Outside the United States, more than 950 cases of H5N1 bird flu have been reported to the World Health Organization. Notably, the Louisiana Department of Health in the United States reported on 6 January 2025 the first U.S. death from a human infection with the H5N1 HPAI virus (www.cdc.gov/media, accessed on 1 April 2025). Dr. James

Lawler, director of the Center for Global Health Security at the University of Nebraska, said the latest changes remind us that "the more widely a virus spreads, especially infecting humans and other mammals, the higher the risk of the virus mutating and adapting to human disease and transmission. That puts all of us at risk".

4. Cross-Species Transmission Mechanism

As mentioned above, cross-species transmission of influenza viruses puts humans at risk. Current research on the cross-species transmission of influenza viruses has highlighted the complex dynamics involved in the adaptation and spread of these viruses among different hosts. The mechanisms underlying this cross-species transmission are multifaceted and involve both viral and host factors. One of the key findings in recent studies is the crucial role of receptor binding specificity in the transmission of influenza A viruses (IAVs). These viruses attach to sialic acid (SA) residues on the surface of host cells, and the type of linkage—either α2,3 or α2,6—plays a pivotal role in determining host susceptibility [10,69]. Avian influenza viruses typically recognize α2,3-linked sialic acids, which are abundant in the intestinal and respiratory tracts of birds, whereas human-adapted strains preferentially bind to α2,6-linked sialic acids, predominantly expressed in the human upper respiratory tract [70,71]. This difference in receptor binding preference represents a major barrier to interspecies transmission. However, accumulating evidence indicates that specific mutations in the viral hemagglutinin (HA) protein, particularly at residues 226 and 228 (H3 numbering), can shift binding specificity from α2,3- to α2,6-linked sialic acids. Notably, substitutions such as Q226L and G228S enhance the virus's ability to recognize and infect human epithelial cells, thereby facilitating more efficient replication and increasing the potential for human-to-human transmission [72–74]. These mutations have been identified in multiple zoonotic strains, including the H2N2 and H3N2 subtypes responsible for past influenza pandemics, highlighting their critical role in host adaptation and cross-species transmission.

Furthermore, researchers conducted surface plasmon resonance (SPR) experiments to detect the virus-encoded hemagglutinin (HA) protein. They discovered that the HA of the bovine H5N1 influenza virus preferentially binds to avian α2-3 sialic acid receptors, while also demonstrating a weak binding affinity to human α2-6 sialic acid receptors. Subsequent immunohistochemical staining indicated that the H5N1 HA protein exhibits strong binding to bovine lung and breast tissues, correlating with the observed clinical symptoms. In contrast, the HA proteins of common human seasonal influenza viruses (H1N1 and H3N2) do not effectively bind to the corresponding tissues in cattle [75]. Moreover, the bovine H5N1 HA protein shows significant binding to human conjunctiva, trachea, lung, and breast tissues, whereas the HA proteins of H1N1 and H3N2 viruses demonstrate significant binding only to human tracheal tissue. This finding provides a comprehensive explanation for the molecular mechanism underlying the association of the bovine H5N1 virus with human conjunctivitis and underscores its distinct tissue tropism compared to seasonal influenza viruses.

The evolutionary dynamics of influenza viruses also play a critical role in cross-species transmission. Over the past century, the emergence of an influenza A virus featuring a new NP gene segment has taken place on just two occasions, each resulting in pandemics: the first in 1918 with the "Spanish" H1N1, which originated from an avian virus, and the second in 2009 with pH1N1, a reassortant virus made up of gene segments from two swine influenza viruses that formed a stable lineage in the human population [34,76,77]. Studies have demonstrated that the introduction of specific mutations, such as the NP-R351K mutation, can significantly enhance the transmissibility of swine influenza viruses. This particular mutation increases viral transmission through multiple mechanisms: enhanced polymerase activity, improved ribonucleoprotein (RNP) assembly, greater thermostability, and more efficient nuclear import. Together, these effects elevate transmission efficiency by 35–40% in mammalian hosts, demonstrating how even minor genetic modifications can substantially impact viral spread [13,78].

In addition to viral factors, host characteristics are also crucial in determining the success of cross-species transmission [79]. The composition of the upper respiratory tract (URT) flora, for example, can influence the susceptibility of hosts to IAV infection [80]. Recent research has utilized mouse models to explore how variations in the URT environment affect viral transmission. It was found that infant mice, which exhibit different sialic acid profiles compared to adults, support more efficient transmission of IAV among littermates [80–83]. This suggests that age and the specific composition of the URT can significantly impact the dynamics of viral spread. Additionally, host factors such as immune status and underlying health conditions can alter the efficiency of influenza virus transmission. For example, children and immunocompromised individuals often exhibit higher viral loads and prolonged periods of viral shedding, thereby increasing the risk of transmission [84,85]. Furthermore, host immune responses, including innate antiviral defenses like interferon signaling, can either limit or facilitate viral propagation depending on their timing and strength [86,87].

Moreover, environmental changes play a crucial role in shaping the transmission dynamics of influenza viruses. Key factors such as temperature, humidity, and seasonal variation directly influence viral stability and infectivity. In temperate regions, influenza activity peaks during colder months, where low temperatures and reduced absolute humidity enhance the stability of influenza virions in aerosols and on surfaces, thereby promoting airborne transmission. Conversely, in tropical regions, transmission patterns are often associated with the rainy season and increased indoor crowding [88,89]. Furthermore, environmental conditions can indirectly affect human behavior, such as increased indoor activity during cold or wet weather, which facilitates close contact and enhances the likelihood of person-to-person transmission [90,91]. Climate change may also alter these patterns by shifting the seasonality of influenza or changing the geographical distribution of outbreaks.

Additionally, environmental factors, such as close contact between hosts and the presence of contaminated surfaces, can facilitate transmission [92]. For instance, the housing conditions of swine farms have been shown to enhance the likelihood of avian viruses spilling over into swine populations, which can then serve as intermediaries for transmission to humans [93].

5. Human Activities and Cross-Species Transmission of Influenza Viruses

Human activities significantly influence the dynamics of cross-species transmission of influenza viruses, primarily through alterations in animal populations, environmental changes, and increased interactions between humans and wildlife [94–97]. The interplay between these factors creates a complex landscape that facilitates the emergence and spread of zoonotic influenza viruses (Figure 3).

One of the most critical aspects of human impact is the increase in livestock populations, particularly poultry and swine, which has been linked to a rise in zoonotic influenza infections [92,98]. The industrialization of agriculture has led to higher densities of animals in confined spaces, creating ideal conditions for the transmission of viruses among species. For instance, the rise in poultry and swine populations since the industrial revolution has been associated with an increased frequency of zoonotic influenza virus infections in humans, highlighting the direct correlation between human agricultural practices and the risk of influenza outbreaks [92,99,100]. Moreover, specific human activities, such as illegal trading of birds, live poultry markets, and practices like cockfighting [101,102], have been identified as significant contributors to the spread of highly pathogenic avian influenza (HPAI) viruses, such as H5N1 [80,103]. These activities not only increase the likelihood of direct contact between infected animals and humans but also facilitate the mixing of different virus strains, which can lead to the emergence of new variants capable of infecting humans [50,103–105].

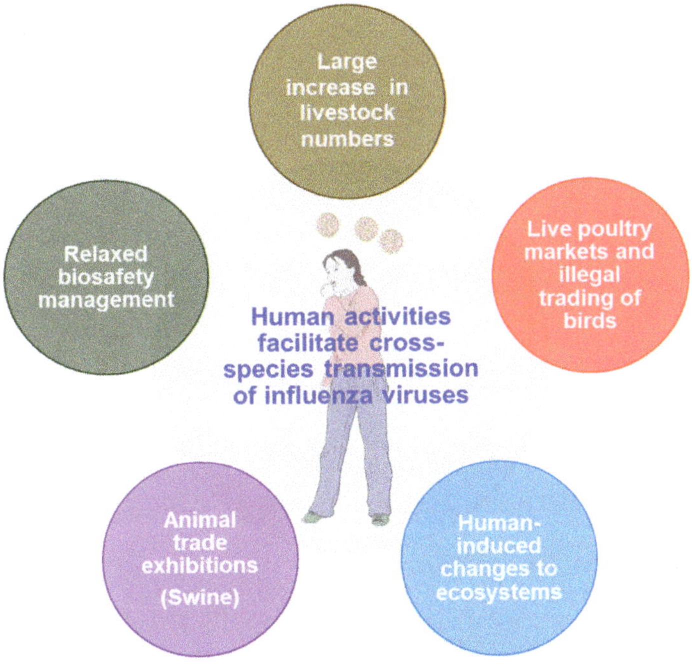

Figure 3. Human activities facilitate cross-species transmission of influenza viruses.

Environmental factors also play a crucial role in cross-species transmission. Human-induced changes to ecosystems, such as deforestation, urbanization, and climate change, can disrupt the natural habitats of wildlife, forcing animals into closer contact with human populations [96,106]. This increased interaction raises the risk of zoonotic spillover events, where viruses jump from animals to humans. For example, the movement of migratory birds, which can carry influenza viruses, is influenced by environmental changes, potentially leading to new transmission pathways [107,108].

Moreover, the exhibition swine industry exemplifies how human management practices can facilitate the rapid dissemination of influenza viruses. Close and prolonged exposure to exhibition swine has led to variant influenza infections in humans, particularly among swine exhibitors [24,49,109]. The concurrent detection of genetically identical influenza A viruses from exhibition swine across different states highlights the mobility of this population and the potential for rapid virus spread [24,110,111]. The relaxed biosecurity in these settings further exacerbates the risk of transmission, allowing for the movement of pathogens across large geographic areas.

6. Strategies and Prospects for Human Response to Interspecies Transmission of Influenza Virus

Cross-species transmission of influenza viruses poses significant challenges to public health, particularly due to the potential for novel strains to emerge that can infect humans and cause pandemics. Addressing this issue requires a multifaceted approach that includes surveillance, biosecurity measures, vaccination strategies, and research into the mechanisms of viral adaptation. One of the primary strategies for managing cross-species transmission is the implementation of robust surveillance systems. Continuous monitoring of influenza viruses in both human and animal populations is essential for the early detection of emerging strains. This includes surveillance of wildlife, domestic animals, and humans, especially in areas where interactions between humans and animals are common. For example, monitoring pig populations is crucial, as pigs can act as mixing vessels for avian and human influenza viruses. This interaction can lead to the emergence of reassortant strains that may pose a risk of pandemics [112,113]. In addition to surveillance, enhancing biosecurity practices in agricultural settings is crucial. This involves improving the separation of different animal species, such as pigs and poultry, to reduce the risk of interspecies transmission. Farmers should adopt better biosecurity measures, such as controlling animal movement, maintaining clean facilities, and minimizing contact between livestock and wild birds [112]. Furthermore, public health strategies should be developed to educate farmers and traders about the risks associated with influenza transmission and the importance of implementing these biosecurity measures.

Vaccination is another vital component in the fight against influenza viruses. Current seasonal influenza vaccines are effective against strains that are closely related to those included in the vaccine; however, the antigenic variability of influenza viruses necessitates annual updates to the vaccine composition [114]. The development of a universal influenza vaccine that can provide broad protection against multiple strains is a promising area of research. Such a vaccine would target conserved viral epitopes, potentially offering protection against a wide range of influenza viruses, including those that have not yet emerged [114].

While human vaccination remains a cornerstone of influenza prevention, controlling influenza viruses at the animal level is equally crucial, particularly in species that serve as reservoirs or intermediate hosts, such as swine and poultry [49,115]. These animals often act as "mixing vessels", where avian, swine, and human influenza strains can reassort, giving rise to novel variants with zoonotic or even pandemic potential [49,116]. Vaccination in these animal populations not only reduces the burden of disease within the species but also significantly decreases viral shedding and transmission risk to humans. For instance, targeted immunization programs in swine have been shown to mitigate the risk of variant influenza virus emergence at agricultural fairs [117,118]. Similarly, vaccination in poultry is a critical control strategy in limiting the spread of highly pathogenic avian influenza and preventing spillover to human populations. Therefore, integrating animal vaccination into a One Health approach is essential for comprehensive influenza control and pandemic preparedness.

Research into the molecular mechanisms of influenza virus adaptation is also essential for understanding cross-species transmission. Studies have shown that specific mutations in viral proteins, such as the polymerase basic protein 2 (PB2), play a significant role in the virus's ability to adapt to new hosts [119]. Understanding these mechanisms can inform the development of targeted interventions to prevent the emergence of new strains. For example, identifying mutations that confer increased transmissibility in mammals can help prioritize surveillance efforts for specific viral variants [83,113,120].

Moreover, the role of environmental factors in facilitating cross-species transmission should not be overlooked. Ecological changes, such as habitat destruction and climate change, can alter the dynamics of host interactions and increase the likelihood of spillover events. Addressing these environmental factors through conservation efforts and sustainable agricultural practices can help mitigate the risk of influenza virus transmission [121].

Finally, international collaboration among public health officials, veterinarians, and researchers is crucial for effectively managing the risks associated with cross-species transmission. Sharing data and resources can enhance the global response to influenza outbreaks and improve preparedness for potential pandemics. Collaborative efforts can also facilitate the development of comprehensive strategies that integrate animal and human health, often referred to as a One Health approach [92]. In summary, addressing the cross-species transmission of influenza viruses necessitates a multifaceted approach that includes surveillance, biosecurity measures, vaccination strategies, research into viral adaptation, and international collaboration (Figure 4). By employing these strategies, public health officials can more effectively manage the risks associated with influenza viruses and diminish the likelihood of future pandemics.

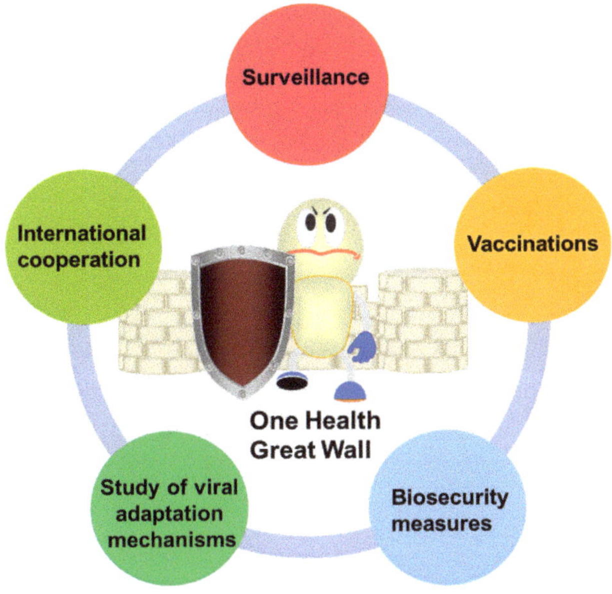

Figure 4. The Great Wall of One Health for all mankind.

7. Conclusions

Cross-species transmission of influenza viruses results from the intricate interplay between viral evolution, host adaptability, and ecological environments. This phenomenon poses significant challenges to global public health security. The essence of cross-species transmission lies in the virus's pursuit of survival advantages within dynamic interactions. Despite notable advancements in understanding the underlying mechanisms and developing intervention strategies, the unpredictability of viral mutations, the complexity of ecosystems, and the impacts of human activities continue to elevate the risk of future pandemics. To effectively address this issue, it is essential to foster continuous innovation in basic research, promote global resource sharing for monitoring, and enhance public health literacy. These efforts will empower us to take the initiative in the ongoing struggle between "viral evolution" and "human defense".

Recent technical advancements have enabled researchers to investigate influenza virus infections across various biological levels, ranging from cellular interactions to whole organisms. This progress has contributed to a deeper understanding of the virus's biology, disease severity, and the immune response. However, despite these advancements, our knowledge remains incomplete regarding the viral properties that could potentially lead to pandemics. Future research should focus on comprehensive mutagenesis studies, cataloging receptor binding, and expanding surveillance in poultry and pigs, particularly in regions such as Africa and South America. These efforts aim to develop robust computational tools and novel animal models to enhance our understanding of transmission dynamics and inform vaccine development strategies [120,122]. It is important that we strive to advance human research proactively in anticipation of viral mutations, rather than relying solely on passive prevention and control measures after a pandemic has emerged.

In summary, the current research status on the cross-species transmission of influenza viruses reveals a dynamic interplay between viral genetics, host factors, and environmental conditions. Understanding these mechanisms is crucial for predicting and mitigating the risks associated with influenza pandemics.

Author Contributions: Conceptualization X.H.; writing—original draft preparation, X.H.; writing—review and editing, X.H., X.T., H.W., C.Z., S.D., L.D., G.G., J.C. and T.Z.; supervision X.H. All authors have read and agreed to the published version of the manuscript.

Funding: This research was supported by the National Natural Science Foundation of China (32300120), the Key Scientific Research Project of Higher Education of Henan Province (25A310013), Natural Science Foundation of Henan Province (232300421315), the 111 Project (D20036, China), Xinxiang Medical University, and the Project for Young Scientists of Henan Province (225200810074). The authors thank all members of the lab for their support.

Institutional Review Board Statement: Not applicable.

Informed Consent Statement: Not applicable.

Data Availability Statement: No new data were created or analyzed in this study. Data sharing is not applicable to this article.

Conflicts of Interest: The authors declare no conflicts of interest.

References

1. Hung, S.J.; Hsu, Y.M.; Huang, S.W.; Tsai, H.P.; Lee, L.Y.Y.; Hurt, A.C.; Barr, I.G.; Shih, S.R.; Wang, J.R. Genetic variations on 31 and 450 residues of influenza A nucleoprotein affect viral replication and translation. *J. Biomed. Sci.* **2020**, *27*, 17. [CrossRef] [PubMed]
2. Dai, M.; McBride, R.; Dortmans, J.; Peng, W.; Bakkers, M.J.G.; de Groot, R.J.; van Kuppeveld, F.J.M.; Paulson, J.C.; de Vries, E.; de Haan, C.A.M. Mutation of the Second Sialic Acid-Binding Site, Resulting in Reduced Neuraminidase Activity, Preceded the Emergence of H7N9 Influenza A Virus. *J. Virol.* **2017**, *91*, e00049-17. [CrossRef] [PubMed]
3. Wang, M.; Zhang, Z.; Wang, X. Strain-Specific Antagonism of the Human H1N1 Influenza A Virus against Equine Tetherin. *Viruses* **2018**, *10*, 264. [CrossRef]
4. Huang, J.; Li, K.; Xiao, S.; Hu, J.; Yin, Y.; Zhang, J.; Li, S.; Wang, W.; Hong, J.; Zhao, Z.; et al. Global epidemiology of animal influenza infections with explicit virus subtypes until 2016: A spatio-temporal descriptive analysis. *One Health* **2023**, *16*, 100514. [CrossRef] [PubMed]
5. Liu, L.; Zeng, X.; Chen, P.; Deng, G.; Li, Y.; Shi, J.; Gu, C.; Kong, H.; Suzuki, Y.; Jiang, Y.; et al. Characterization of Clade 7.2 H5 Avian Influenza Viruses That Continue To Circulate in Chickens in China. *J. Virol.* **2016**, *90*, 9797–9805. [CrossRef]
6. Liang, L.; Jiang, L.; Li, J.; Zhao, Q.; Wang, J.; He, X.; Huang, S.; Wang, Q.; Zhao, Y.; Wang, G.; et al. Low Polymerase Activity Attributed to PA Drives the Acquisition of the PB2 E627K Mutation of H7N9 Avian Influenza Virus in Mammals. *mBio* **2019**, *10*, e01162-19. [CrossRef]
7. Shi, Y.; Zhang, W.; Wang, F.; Qi, J.; Wu, Y.; Song, H.; Gao, F.; Bi, Y.; Zhang, Y.; Fan, Z.; et al. Structures and receptor binding of hemagglutinins from human-infecting H7N9 influenza viruses. *Science* **2013**, *342*, 243–247. [CrossRef]
8. Xiong, X.; Martin, S.R.; Haire, L.F.; Wharton, S.A.; Daniels, R.S.; Bennett, M.S.; McCauley, J.W.; Collins, P.J.; Walker, P.A.; Skehel, J.J.; et al. Receptor binding by an H7N9 influenza virus from humans. *Nature* **2013**, *499*, 496–499. [CrossRef]
9. de Vries, R.P.; Peng, W.; Grant, O.C.; Thompson, A.J.; Zhu, X.; Bouwman, K.M.; de la Pena, A.T.T.; van Breemen, M.J.; Ambepitiya Wickramasinghe, I.N.; de Haan, C.A.M.; et al. Three mutations switch H7N9 influenza to human-type receptor specificity. *PLoS Pathog.* **2017**, *13*, e1006390. [CrossRef]
10. Long, J.S.; Mistry, B.; Haslam, S.M.; Barclay, W.S. Host and viral determinants of influenza A virus species specificity. *Nat. Rev. Microbiol.* **2019**, *17*, 67–81. [CrossRef]

11. Garcia-Sastre, A. Inhibition of interferon-mediated antiviral responses by influenza A viruses and other negative-strand RNA viruses. *Virology* **2001**, *279*, 375–384. [CrossRef] [PubMed]
12. Hale, B.G.; Randall, R.E.; Ortin, J.; Jackson, D. The multifunctional NS1 protein of influenza A viruses. *J. Gen. Virol.* **2008**, *89 Pt 10*, 2359–2376. [CrossRef]
13. Mehle, A.; Doudna, J.A. Adaptive strategies of the influenza virus polymerase for replication in humans. *Proc. Natl. Acad. Sci. USA* **2009**, *106*, 21312–21316. [CrossRef]
14. Steel, J.; Lowen, A.C.; Mubareka, S.; Palese, P. Transmission of influenza virus in a mammalian host is increased by PB2 amino acids 627K or 627E/701N. *PLoS Pathog.* **2009**, *5*, e1000252. [CrossRef]
15. Osterhaus, A.D.; Rimmelzwaan, G.F.; Martina, B.E.; Bestebroer, T.M.; Fouchier, R.A. Influenza B virus in seals. *Science* **2000**, *288*, 1051–1053. [CrossRef] [PubMed]
16. Krammer, F.; Smith, G.J.D.; Fouchier, R.A.M.; Peiris, M.; Kedzierska, K.; Doherty, P.C.; Palese, P.; Shaw, M.L.; Treanor, J.; Webster, R.G.; et al. Influenza. *Nat. Rev. Dis. Primers* **2018**, *4*, 3. [CrossRef] [PubMed]
17. Anthony, S.J.; Epstein, J.H.; Murray, K.A.; Navarrete-Macias, I.; Zambrana-Torrelio, C.M.; Solovyov, A.; Ojeda-Flores, R.; Arrigo, N.C.; Islam, A.; Ali Khan, S.; et al. A strategy to estimate unknown viral diversity in mammals. *mBio* **2013**, *4*, e00598-13. [CrossRef]
18. Olsen, B.; Munster, V.J.; Wallensten, A.; Waldenstrom, J.; Osterhaus, A.D.; Fouchier, R.A. Global patterns of influenza a virus in wild birds. *Science* **2006**, *312*, 384–388. [CrossRef]
19. Wille, M.; Holmes, E.C. The Ecology and Evolution of Influenza Viruses. *Cold Spring Harb. Perspect. Med.* **2020**, *10*, a038489. [CrossRef]
20. de Graaf, M.; Fouchier, R.A. Role of receptor binding specificity in influenza A virus transmission and pathogenesis. *EMBO J.* **2014**, *33*, 823–841. [CrossRef]
21. Taubenberger, J.K.; Kash, J.C. Influenza virus evolution, host adaptation, and pandemic formation. *Cell Host Microbe* **2010**, *7*, 440–451. [CrossRef] [PubMed]
22. Karesh, W.B.; Dobson, A.; Lloyd-Smith, J.O.; Lubroth, J.; Dixon, M.A.; Bennett, M.; Aldrich, S.; Harrington, T.; Formenty, P.; Loh, E.H.; et al. Ecology of zoonoses: Natural and unnatural histories. *Lancet* **2012**, *380*, 1936–1945. [CrossRef] [PubMed]
23. Nelson, M.I.; Wentworth, D.E.; Das, S.R.; Sreevatsan, S.; Killian, M.L.; Nolting, J.M.; Slemons, R.D.; Bowman, A.S. Evolutionary Dynamics of Influenza A Viruses in US Exhibition Swine. *J. Infect. Dis.* **2016**, *213*, 173–182. [CrossRef] [PubMed]
24. Bowman, A.S.; Walia, R.R.; Nolting, J.M.; Vincent, A.L.; Killian, M.L.; Zentkovich, M.M.; Lorbach, J.N.; Lauterbach, S.E.; Anderson, T.K.; Davis, C.T.; et al. Influenza A(H3N2) Virus in Swine at Agricultural Fairs and Transmission to Humans, Michigan and Ohio, USA, 2016. *Emerg. Infect. Dis.* **2017**, *23*, 1551–1555. [CrossRef]
25. Fobian, K.; Fabrizio, T.P.; Yoon, S.W.; Hansen, M.S.; Webby, R.J.; Larsen, L.E. New reassortant and enzootic European swine influenza viruses transmit efficiently through direct contact in the ferret model. *J. Gen. Virol.* **2015**, *96*, 1603–1612. [CrossRef]
26. Hause, B.M.; Collin, E.A.; Liu, R.; Huang, B.; Sheng, Z.; Lu, W.; Wang, D.; Nelson, E.A.; Li, F. Characterization of a novel influenza virus in cattle and Swine: Proposal for a new genus in the Orthomyxoviridae family. *mBio* **2014**, *5*, e00031-14. [CrossRef]
27. Hause, B.M.; Ducatez, M.; Collin, E.A.; Ran, Z.; Liu, R.; Sheng, Z.; Armien, A.; Kaplan, B.; Chakravarty, S.; Hoppe, A.D.; et al. Isolation of a novel swine influenza virus from Oklahoma in 2011 which is distantly related to human influenza C viruses. *PLoS Pathog.* **2013**, *9*, e1003176. [CrossRef]
28. Chiapponi, C.; Faccini, S.; De Mattia, A.; Baioni, L.; Barbieri, I.; Rosignoli, C.; Nigrelli, A.; Foni, E. Detection of Influenza D Virus among Swine and Cattle, Italy. *Emerg. Infect. Dis.* **2016**, *22*, 352–354. [CrossRef]
29. Cox, N.J.; Subbarao, K. Global epidemiology of influenza: Past and present. *Annu. Rev. Med.* **2000**, *51*, 407–421. [CrossRef]
30. Kobasa, D.; Takada, A.; Shinya, K.; Hatta, M.; Halfmann, P.; Theriault, S.; Suzuki, H.; Nishimura, H.; Mitamura, K.; Sugaya, N.; et al. Enhanced virulence of influenza A viruses with the haemagglutinin of the 1918 pandemic virus. *Nature* **2004**, *431*, 703–707. [CrossRef]
31. Novel Swine-Origin Influenza A (H1N1) Virus Investigation Team; Dawood, F.S.; Jain, S.; Finelli, L.; Shaw, M.W.; Lindstrom, S.; Garten, R.J.; Gubareva, L.V.; Xu, X.; Bridges, C.B.; et al. Emergence of a novel swine-origin influenza A (H1N1) virus in humans. *N. Engl. J. Med.* **2009**, *360*, 2605–2615. [CrossRef]
32. Bi, Y.; Yang, J.; Wang, L.; Ran, L.; Gao, G.F. Ecology and evolution of avian influenza viruses. *Curr. Biol.* **2024**, *34*, R716–R721. [CrossRef]
33. Ito, T.; Couceiro, J.N.; Kelm, S.; Baum, L.G.; Krauss, S.; Castrucci, M.R.; Donatelli, I.; Kida, H.; Paulson, J.C.; Webster, R.G.; et al. Molecular basis for the generation in pigs of influenza A viruses with pandemic potential. *J. Virol.* **1998**, *72*, 7367–7373. [CrossRef] [PubMed]
34. Smith, G.J.; Vijaykrishna, D.; Bahl, J.; Lycett, S.J.; Worobey, M.; Pybus, O.G.; Ma, S.K.; Cheung, C.L.; Raghwani, J.; Bhatt, S.; et al. Origins and evolutionary genomics of the 2009 swine-origin H1N1 influenza A epidemic. *Nature* **2009**, *459*, 1122–1125. [CrossRef]
35. Paul Glezen, W.; Schmier, J.K.; Kuehn, C.M.; Ryan, K.J.; Oxford, J. The burden of influenza B: A structured literature review. *Am. J. Public Health* **2013**, *103*, e43–e51. [CrossRef] [PubMed]
36. Glezen, W.P. Editorial commentary: Changing epidemiology of influenza B virus. *Clin. Infect. Dis.* **2014**, *59*, 1525–1526. [CrossRef]

37. Abdullahi, O.; Karani, A.; Tigoi, C.C.; Mugo, D.; Kungu, S.; Wanjiru, E.; Jomo, J.; Musyimi, R.; Lipsitch, M.; Scott, J.A. Rates of acquisition and clearance of pneumococcal serotypes in the nasopharynges of children in Kilifi District, Kenya. *J. Infect. Dis.* **2012**, *206*, 1020–1029. [CrossRef]
38. Belser, J.A.; Eckert, A.M.; Tumpey, T.M.; Maines, T.R. Complexities in Ferret Influenza Virus Pathogenesis and Transmission Models. *Microbiol. Mol. Biol. Rev.* **2016**, *80*, 733–744. [CrossRef]
39. Laurie, K.L.; Carolan, L.A.; Middleton, D.; Lowther, S.; Kelso, A.; Barr, I.G. Multiple infections with seasonal influenza A virus induce cross-protective immunity against A(H1N1) pandemic influenza virus in a ferret model. *J. Infect. Dis.* **2010**, *202*, 1011–1020. [CrossRef]
40. Chen, R.; Holmes, E.C. The evolutionary dynamics of human influenza B virus. *J. Mol. Evol.* **2008**, *66*, 655–663. [CrossRef]
41. Langat, P.; Raghwani, J.; Dudas, G.; Bowden, T.A.; Edwards, S.; Gall, A.; Bedford, T.; Rambaut, A.; Daniels, R.S.; Russell, C.A.; et al. Genome-wide evolutionary dynamics of influenza B viruses on a global scale. *PLoS Pathog.* **2017**, *13*, e1006749. [CrossRef]
42. Vijaykrishna, D.; Holmes, E.C.; Joseph, U.; Fourment, M.; Su, Y.C.; Halpin, R.; Lee, R.T.; Deng, Y.M.; Gunalan, V.; Lin, X.; et al. The contrasting phylodynamics of human influenza B viruses. *eLife* **2015**, *4*, e05055. [CrossRef]
43. Liu, R.; Sheng, Z.; Lin, T.; Sreenivasan, C.; Gao, R.; Thomas, M.; Druce, J.; Hause, B.M.; Kaushik, R.S.; Li, F.; et al. Genetic and antigenic characteristics of a human influenza C virus clinical isolate. *J. Med. Virol.* **2020**, *92*, 161–166. [CrossRef]
44. Sederdahl, B.K.; Williams, J.V. Epidemiology and Clinical Characteristics of Influenza C Virus. *Viruses* **2020**, *12*, 89. [CrossRef]
45. Kwasnik, M.; Rola, J.; Rozek, W. Influenza D in Domestic and Wild Animals. *Viruses* **2023**, *15*, 2433. [CrossRef] [PubMed]
46. Sreenivasan, C.C.; Liu, R.; Gao, R.; Guo, Y.; Hause, B.M.; Thomas, M.; Naveed, A.; Clement, T.; Rausch, D.; Christopher-Hennings, J.; et al. Influenza C and D Viruses Demonstrated a Differential Respiratory Tissue Tropism in a Comparative Pathogenesis Study in Guinea Pigs. *J. Virol.* **2023**, *97*, e0035623. [CrossRef] [PubMed]
47. Matsuzaki, Y.; Katsushima, N.; Nagai, Y.; Shoji, M.; Itagaki, T.; Sakamoto, M.; Kitaoka, S.; Mizuta, K.; Nishimura, H. Clinical features of influenza C virus infection in children. *J. Infect. Dis.* **2006**, *193*, 1229–1235. [CrossRef] [PubMed]
48. Su, S.; Fu, X.; Li, G.; Kerlin, F.; Veit, M. Novel Influenza D virus: Epidemiology, pathology, evolution and biological characteristics. *Virulence* **2017**, *8*, 1580–1591. [CrossRef]
49. Nelson, M.I.; Vincent, A.L. Reverse zoonosis of influenza to swine: New perspectives on the human-animal interface. *Trends Microbiol.* **2015**, *23*, 142–153. [CrossRef]
50. Webster, R.G.; Bean, W.J.; Gorman, O.T.; Chambers, T.M.; Kawaoka, Y. Evolution and ecology of influenza A viruses. *Microbiol. Rev.* **1992**, *56*, 152–179. [CrossRef]
51. Hui, X.; Cao, L.; Xu, T.; Zhao, L.; Huang, K.; Zou, Z.; Ren, P.; Mao, H.; Yang, Y.; Gao, S.; et al. PSMD12-Mediated M1 Ubiquitination of Influenza A Virus at K102 Regulates Viral Replication. *J. Virol.* **2022**, *96*, e0078622. [CrossRef] [PubMed]
52. Taubenberger, J.K.; Morens, D.M. 1918 Influenza: The mother of all pandemics. *Emerg. Infect. Dis.* **2006**, *12*, 15–22. [CrossRef]
53. Worobey, M.; Han, G.Z.; Rambaut, A. Genesis and pathogenesis of the 1918 pandemic H1N1 influenza A virus. *Proc. Natl. Acad. Sci. USA* **2014**, *111*, 8107–8112. [CrossRef] [PubMed]
54. Kawaoka, Y.; Krauss, S.; Webster, R.G. Avian-to-human transmission of the PB1 gene of influenza A viruses in the 1957 and 1968 pandemics. *J. Virol.* **1989**, *63*, 4603–4608. [CrossRef]
55. Scholtissek, C.; Rohde, W.; Von Hoyningen, V.; Rott, R. On the origin of the human influenza virus subtypes H_2N_2 and H_3N_2. *Virology* **1978**, *87*, 13–20. [CrossRef] [PubMed]
56. Smith, G.J.; Bahl, J.; Vijaykrishna, D.; Zhang, J.; Poon, L.L.; Chen, H.; Webster, R.G.; Peiris, J.S.; Guan, Y. Dating the emergence of pandemic influenza viruses. *Proc. Natl. Acad. Sci. USA* **2009**, *106*, 11709–11712. [CrossRef]
57. Garten, R.J.; Davis, C.T.; Russell, C.A.; Shu, B.; Lindstrom, S.; Balish, A.; Sessions, W.M.; Xu, X.; Skepner, E.; Deyde, V.; et al. Antigenic and genetic characteristics of swine-origin 2009 A(H1N1) influenza viruses circulating in humans. *Science* **2009**, *325*, 197–201. [CrossRef]
58. Horimoto, T.; Kawaoka, Y. Pandemic threat posed by avian influenza A viruses. *Clin. Microbiol. Rev.* **2001**, *14*, 129–149. [CrossRef]
59. Tundup, S.; Kandasamy, M.; Perez, J.T.; Mena, N.; Steel, J.; Nagy, T.; Albrecht, R.A.; Manicassamy, B. Endothelial cell tropism is a determinant of H5N1 pathogenesis in mammalian species. *PLoS Pathog.* **2017**, *13*, e1006270. [CrossRef]
60. Horimoto, T.; Kawaoka, Y. Influenza: Lessons from past pandemics, warnings from current incidents. *Nat. Rev. Microbiol.* **2005**, *3*, 591–600. [CrossRef]
61. Dibarbora, M.; Cappuccio, J.; Olivera, V.; Quiroga, M.; Machuca, M.; Perfumo, C.; Perez, D.; Pereda, A. Swine influenza: Clinical, serological, pathological, and virological cross-sectional studies in nine farms in Argentina. *Influenza Other Respir. Viruses* **2013**, *7* (Suppl. S4), 10–15. [CrossRef] [PubMed]
62. Alexander, D.J. A review of avian influenza in different bird species. *Vet. Microbiol.* **2000**, *74*, 3–13. [CrossRef]
63. Song, Y.; Li, W.; Wu, W.; Liu, Z.; He, Z.; Chen, Z.; Zhao, B.; Wu, S.; Yang, C.; Qu, X.; et al. Phylogeny, Pathogenicity, Transmission, and Host Immune Responses of Four H5N6 Avian Influenza Viruses in Chickens and Mice. *Viruses* **2019**, *11*, 1048. [CrossRef]

64. Chrzastek, K.; Segovia, K.; Torchetti, M.; Killian, M.L.; Pantin-Jackwood, M.; Kapczynski, D.R. Virus Adaptation Following Experimental Infection of Chickens with a Domestic Duck Low Pathogenic Avian Influenza Isolate from the 2017 USA H7N9 Outbreak Identifies Polymorphic Mutations in Multiple Gene Segments. *Viruses* **2021**, *13*, 1166. [CrossRef]
65. Claas, E.C.; Osterhaus, A.D.; van Beek, R.; De Jong, J.C.; Rimmelzwaan, G.F.; Senne, D.A.; Krauss, S.; Shortridge, K.F.; Webster, R.G. Human influenza A H5N1 virus related to a highly pathogenic avian influenza virus. *Lancet* **1998**, *351*, 472–477. [CrossRef] [PubMed]
66. European Food Safety Authority; European Centre for Disease Prevention and Control; European Union Reference Laboratory for Avian Influenza; Adlhoch, C.; Fusaro, A.; Gonzales, J.L.; Kuiken, T.; Marangon, S.; Mirinaviciute, G.; Niqueux, É.; et al. Avian influenza overview December 2022–March 2023. *EFSA J.* **2023**, *21*, e07917. [CrossRef]
67. Yamaji, R.; Zhang, W.; Kamata, A.; Adlhoch, C.; Swayne, D.E.; Pereyaslov, D.; Wang, D.; Neumann, G.; Pavade, G.; Barr, I.G.; et al. Pandemic risk characterisation of zoonotic influenza A viruses using the Tool for Influenza Pandemic Risk Assessment (TIPRA). *Lancet Microbe* **2024**, *6*, 100973. [CrossRef] [PubMed]
68. Song, H.; Hao, T.; Han, P.; Wang, H.; Zhang, X.; Li, X.; Wang, Y.; Chen, J.; Li, Y.; Jin, X.; et al. Receptor binding, structure, and tissue tropism of cattle-infecting H5N1 avian influenza virus hemagglutinin. *Cell* **2025**, *188*, 919–929.e9. [CrossRef]
69. Spruit, C.M.; Zhu, X.; Tomris, I.; Rios-Carrasco, M.; Han, A.X.; Broszeit, F.; van der Woude, R.; Bouwman, K.M.; Luu, M.M.T.; Matsuno, K.; et al. N-Glycolylneuraminic Acid Binding of Avian and Equine H7 Influenza A Viruses. *J. Virol.* **2022**, *96*, e0212021. [CrossRef]
70. Shinya, K.; Ebina, M.; Yamada, S.; Ono, M.; Kasai, N.; Kawaoka, Y. Avian flu: Influenza virus receptors in the human airway. *Nature* **2006**, *440*, 435–436. [CrossRef]
71. van Riel, D.; Munster, V.J.; de Wit, E.; Rimmelzwaan, G.F.; Fouchier, R.A.; Osterhaus, A.D.; Kuiken, T. H5N1 Virus Attachment to Lower Respiratory Tract. *Science* **2006**, *312*, 399. [CrossRef] [PubMed]
72. Rios Carrasco, M.; Lin, T.H.; Zhu, X.; Garcia, A.G.; Uslu, E.; Liang, R.; Spruit, C.M.; Richard, M.; Boons, G.J.; Wilson, I.A.; et al. The Q226L mutation can convert a highly pathogenic H5 2.3.4.4e virus to bind human-type receptors. *bioRxiv* **2025**. [CrossRef]
73. Van Poucke, S.; Doedt, J.; Baumann, J.; Qiu, Y.; Matrosovich, T.; Klenk, H.D.; Van Reeth, K.; Matrosovich, M. Role of Substitutions in the Hemagglutinin in the Emergence of the 1968 Pandemic Influenza Virus. *J. Virol.* **2015**, *89*, 12211–12216. [CrossRef]
74. Guo, X.; Zhou, Y.; Yan, H.; An, Q.; Liang, C.; Liu, L.; Qian, J. Molecular Markers and Mechanisms of Influenza A Virus Cross-Species Transmission and New Host Adaptation. *Viruses* **2024**, *16*, 883. [CrossRef] [PubMed]
75. Lin, T.H.; Zhu, X.; Wang, S.; Zhang, D.; McBride, R.; Yu, W.; Babarinde, S.; Paulson, J.C.; Wilson, I.A. A single mutation in bovine influenza H5N1 hemagglutinin switches specificity to human receptors. *Science* **2024**, *386*, 1128–1134. [CrossRef]
76. Taubenberger, J.K.; Kash, J.C. Insights on influenza pathogenesis from the grave. *Virus Res.* **2011**, *162*, 2–7. [CrossRef] [PubMed]
77. Neumann, G.; Noda, T.; Kawaoka, Y. Emergence and pandemic potential of swine-origin H1N1 influenza virus. *Nature* **2009**, *459*, 931–939. [CrossRef]
78. Gabriel, G.; Dauber, B.; Wolff, T.; Planz, O.; Klenk, H.D.; Stech, J. The viral polymerase mediates adaptation of an avian influenza virus to a mammalian host. *Proc. Natl. Acad. Sci. USA* **2005**, *102*, 18590–18595. [CrossRef]
79. Ganti, K.; Bagga, A.; Carnaccini, S.; Ferreri, L.M.; Geiger, G.; Joaquin Caceres, C.; Seibert, B.; Li, Y.; Wang, L.; Kwon, T.; et al. Influenza A virus reassortment in mammals gives rise to genetically distinct within-host subpopulations. *Nat. Commun.* **2022**, *13*, 6846. [CrossRef]
80. Ortigoza, M.B.; Blaser, S.B.; Zafar, M.A.; Hammond, A.J.; Weiser, J.N. An Infant Mouse Model of Influenza Virus Transmission Demonstrates the Role of Virus-Specific Shedding, Humoral Immunity, and Sialidase Expression by Colonizing Streptococcus pneumoniae. *mBio* **2018**, *9*, e02359-18. [CrossRef]
81. Eaton, M.D. Transmission of Epidemic Influenza Virus in Mice by Contact. *J. Bacteriol.* **1940**, *39*, 229–241. [CrossRef]
82. Diavatopoulos, D.A.; Short, K.R.; Price, J.T.; Wilksch, J.J.; Brown, L.E.; Briles, D.E.; Strugnell, R.A.; Wijburg, O.L. Influenza A virus facilitates Streptococcus pneumoniae transmission and disease. *FASEB J.* **2010**, *24*, 1789–1798. [CrossRef] [PubMed]
83. Luan, H. Cell-Autonomous and Non-Cell-Autonomous Antiviral Immunity via siRNA-Directed RNAi in Drosophila melanogaster. *Immune Discov.* **2025**, *1*, 10001. [CrossRef]
84. Ip, D.K.; Lau, L.L.; Leung, N.H.; Fang, V.J.; Chan, K.H.; Chu, D.K.; Leung, G.M.; Peiris, J.S.; Uyeki, T.M.; Cowling, B.J. Viral Shedding and Transmission Potential of Asymptomatic and Paucisymptomatic Influenza Virus Infections in the Community. *Clin. Infect. Dis.* **2017**, *64*, 736–742. [CrossRef] [PubMed]
85. Memoli, M.J.; Czajkowski, L.; Reed, S.; Athota, R.; Bristol, T.; Proudfoot, K.; Fargis, S.; Stein, M.; Dunfee, R.L.; Shaw, P.A.; et al. Validation of the wild-type influenza A human challenge model H1N1pdMIST: An A(H1N1)pdm09 dose-finding investigational new drug study. *Clin. Infect. Dis.* **2015**, *60*, 693–702. [CrossRef]
86. Iwasaki, A.; Pillai, P.S. Innate immunity to influenza virus infection. *Nat. Rev. Immunol.* **2014**, *14*, 315–328. [CrossRef] [PubMed]
87. Kallfass, C.; Lienenklaus, S.; Weiss, S.; Staeheli, P. Visualizing the beta interferon response in mice during infection with influenza A viruses expressing or lacking nonstructural protein 1. *J. Virol.* **2013**, *87*, 6925–6930. [CrossRef]

88. Shaman, J.; Kohn, M. Absolute humidity modulates influenza survival, transmission, and seasonality. *Proc. Natl. Acad. Sci. USA* **2009**, *106*, 3243–3248. [CrossRef]
89. Lowen, A.C.; Mubareka, S.; Steel, J.; Palese, P. Influenza virus transmission is dependent on relative humidity and temperature. *PLoS Pathog.* **2007**, *3*, 1470–1476. [CrossRef]
90. Tamerius, J.D.; Shaman, J.; Alonso, W.J.; Bloom-Feshbach, K.; Uejio, C.K.; Comrie, A.; Viboud, C. Environmental predictors of seasonal influenza epidemics across temperate and tropical climates. *PLoS Pathog.* **2013**, *9*, e1003194. [CrossRef]
91. Lowen, A.C.; Steel, J. Roles of humidity and temperature in shaping influenza seasonality. *J. Virol.* **2014**, *88*, 7692–7695. [CrossRef] [PubMed]
92. Short, K.R.; Richard, M.; Verhagen, J.H.; van Riel, D.; Schrauwen, E.J.; van den Brand, J.M.; Manz, B.; Bodewes, R.; Herfst, S. One health, multiple challenges: The inter-species transmission of influenza A virus. *One Health* **2015**, *1*, 1–13. [CrossRef]
93. Ma, J.; Shen, H.; Liu, Q.; Bawa, B.; Qi, W.; Duff, M.; Lang, Y.; Lee, J.; Yu, H.; Bai, J.; et al. Pathogenicity and transmissibility of novel reassortant H3N2 influenza viruses with 2009 pandemic H1N1 genes in pigs. *J. Virol.* **2015**, *89*, 2831–2841. [CrossRef] [PubMed]
94. Parrish, C.R.; Holmes, E.C.; Morens, D.M.; Park, E.C.; Burke, D.S.; Calisher, C.H.; Laughlin, C.A.; Saif, L.J.; Daszak, P. Cross-species virus transmission and the emergence of new epidemic diseases. *Microbiol. Mol. Biol. Rev.* **2008**, *72*, 457–470. [CrossRef] [PubMed]
95. Jones, K.E.; Patel, N.G.; Levy, M.A.; Storeygard, A.; Balk, D.; Gittleman, J.L.; Daszak, P. Global trends in emerging infectious diseases. *Nature* **2008**, *451*, 990–993. [CrossRef]
96. Plowright, R.K.; Parrish, C.R.; McCallum, H.; Hudson, P.J.; Ko, A.I.; Graham, A.L.; Lloyd-Smith, J.O. Pathways to zoonotic spillover. *Nat. Rev. Microbiol.* **2017**, *15*, 502–510. [CrossRef]
97. Webster, R.G.; Govorkova, E.A. H5N1 influenza–continuing evolution and spread. *N. Engl. J. Med.* **2006**, *355*, 2174–2177. [CrossRef]
98. Gilbert, M.; Xiao, X.; Pfeiffer, D.U.; Epprecht, M.; Boles, S.; Czarnecki, C.; Chaitaweesub, P.; Kalpravidh, W.; Minh, P.Q.; Otte, M.J.; et al. Mapping H5N1 highly pathogenic avian influenza risk in Southeast Asia. *Proc. Natl. Acad. Sci. USA* **2008**, *105*, 4769–4774. [CrossRef]
99. Greger, M. The human/animal interface: Emergence and resurgence of zoonotic infectious diseases. *Crit. Rev. Microbiol.* **2007**, *33*, 243–299. [CrossRef]
100. Webster, R.G. Wet markets--a continuing source of severe acute respiratory syndrome and influenza? *Lancet* **2004**, *363*, 234–236. [CrossRef]
101. Gilbert, M.; Golding, N.; Zhou, H.; Wint, G.R.; Robinson, T.P.; Tatem, A.J.; Lai, S.; Zhou, S.; Jiang, H.; Guo, D.; et al. Predicting the risk of avian influenza A H7N9 infection in live-poultry markets across Asia. *Nat. Commun.* **2014**, *5*, 4116. [CrossRef] [PubMed]
102. Kung, N.Y.; Morris, R.S.; Perkins, N.R.; Sims, L.D.; Ellis, T.M.; Bissett, L.; Chow, M.; Shortridge, K.F.; Guan, Y.; Peiris, M.J. Risk for infection with highly pathogenic influenza A virus (H5N1) in chickens, Hong Kong, 2002. *Emerg. Infect. Dis.* **2007**, *13*, 412–418. [CrossRef] [PubMed]
103. Peiris, J.S.; de Jong, M.D.; Guan, Y. Avian influenza virus (H5N1): A threat to human health. *Clin. Microbiol. Rev.* **2007**, *20*, 243–267. [CrossRef]
104. Bahl, J.; Pham, T.T.; Hill, N.J.; Hussein, I.T.; Ma, E.J.; Easterday, B.C.; Halpin, R.A.; Stockwell, T.B.; Wentworth, D.E.; Kayali, G.; et al. Ecosystem Interactions Underlie the Spread of Avian Influenza A Viruses with Pandemic Potential. *PLoS Pathog.* **2016**, *12*, e1005620. [CrossRef]
105. Guan, Y.; Poon, L.L.; Cheung, C.Y.; Ellis, T.M.; Lim, W.; Lipatov, A.S.; Chan, K.H.; Sturm-Ramirez, K.M.; Cheung, C.L.; Leung, Y.H.; et al. H5N1 influenza: A protean pandemic threat. *Proc. Natl. Acad. Sci. USA* **2004**, *101*, 8156–8161. [CrossRef] [PubMed]
106. Khan, A.; Ahmed, H.; Amjad, S.; Afzal, M.S.; Haider, W.; Simsek, S.; Khawaja, M.R.; Khan, D.H.; Naz, S.; Durrance-Bagale, A.; et al. Community Based Assessment of Behavior and Awareness of Risk Factors of Cystic Echinococcosis in Major Cities of Pakistan: A One Health Perspective. *Front. Public Health* **2021**, *9*, 648900. [CrossRef]
107. Gilbert, M.; Chaitaweesub, P.; Parakamawongsa, T.; Premashthira, S.; Tiensin, T.; Kalpravidh, W.; Wagner, H.; Slingenbergh, J. Free-grazing ducks and highly pathogenic avian influenza, Thailand. *Emerg. Infect. Dis.* **2006**, *12*, 227–234. [CrossRef]
108. Fuller, T.L.; Gilbert, M.; Martin, V.; Cappelle, J.; Hosseini, P.; Njabo, K.Y.; Abdel Aziz, S.; Xiao, X.; Daszak, P.; Smith, T.B. Predicting hotspots for influenza virus reassortment. *Emerg. Infect. Dis.* **2013**, *19*, 581–588. [CrossRef]
109. Bowman, A.S.; Nelson, S.W.; Page, S.L.; Nolting, J.M.; Killian, M.L.; Sreevatsan, S.; Slemons, R.D. Swine-to-human transmission of influenza A(H3N2) virus at agricultural fairs, Ohio, USA, 2012. *Emerg. Infect. Dis.* **2014**, *20*, 1472–1480. [CrossRef]
110. McBride, D.S.; Nolting, J.M.; Nelson, S.W.; Spurck, M.M.; Bliss, N.T.; Kenah, E.; Trock, S.C.; Bowman, A.S. Shortening Duration of Swine Exhibitions to Reduce Risk for Zoonotic Transmission of Influenza A Virus. *Emerg. Infect. Dis.* **2022**, *28*, 2035–2042. [CrossRef]
111. Nelson, M.I.; Stratton, J.; Killian, M.L.; Janas-Martindale, A.; Vincent, A.L. Continual Reintroduction of Human Pandemic H1N1 Influenza A Viruses into Swine in the United States, 2009 to 2014. *J. Virol.* **2015**, *89*, 6218–6226. [CrossRef] [PubMed]

112. Netrabukkana, P.; Robertson, I.D.; Kasemsuwan, S.; Wongsathapornchai, K.; Fenwick, S. Assessing Potential Risks of Influenza A Virus Transmission at the Pig-Human Interface in Thai Small Pig Farms Using a Questionnaire Survey. *Transbound. Emerg. Dis.* **2016**, *63*, e135–e139. [CrossRef]
113. Lipsitch, M.; Barclay, W.; Raman, R.; Russell, C.J.; Belser, J.A.; Cobey, S.; Kasson, P.M.; Lloyd-Smith, J.O.; Maurer-Stroh, S.; Riley, S.; et al. Viral factors in influenza pandemic risk assessment. *eLife* **2016**, *5*, e18491. [CrossRef] [PubMed]
114. Rudraraju, R.; Subbarao, K. Passive immunization with influenza haemagglutinin specific monoclonal antibodies. *Hum. Vaccin. Immunother.* **2018**, *14*, 2728–2736. [CrossRef] [PubMed]
115. Sims, L.D. Progress in control of H5N1 highly pathogenic avian influenza and the future for eradication. *Avian Dis.* **2012**, *56*, 829–835. [CrossRef]
116. Scholtissek, C.; Burger, H.; Kistner, O.; Shortridge, K.F. The nucleoprotein as a possible major factor in determining host specificity of influenza H3N2 viruses. *Virology* **1985**, *147*, 287–294. [CrossRef]
117. Nelson, M.I.; Wentworth, D.E.; Culhane, M.R.; Vincent, A.L.; Viboud, C.; LaPointe, M.P.; Lin, X.; Holmes, E.C.; Detmer, S.E. Introductions and evolution of human-origin seasonal influenza a viruses in multinational swine populations. *J. Virol.* **2014**, *88*, 10110–10119. [CrossRef] [PubMed]
118. Szablewski, C.M.; McBride, D.S.; Trock, S.C.; Habing, G.G.; Hoet, A.E.; Nelson, S.W.; Nolting, J.M.; Bowman, A.S. Evolution of influenza A viruses in exhibition swine and transmission to humans, 2013–2015. *Zoonoses Public Health* **2024**, *71*, 281–293. [CrossRef]
119. Liu, X.; Yang, C.; Sun, X.; Lin, X.; Zhao, L.; Chen, H.; Jin, M. Evidence for a novel mechanism of influenza A virus host adaptation modulated by PB2-627. *FEBS J.* **2019**, *286*, 3389–3400. [CrossRef]
120. Neumann, G.; Kawaoka, Y. Predicting the Next Influenza Pandemics. *J. Infect. Dis.* **2019**, *219*, S14–S20. [CrossRef]
121. Durand, L.O.; Glew, P.; Gross, D.; Kasper, M.; Trock, S.; Kim, I.K.; Bresee, J.S.; Donis, R.; Uyeki, T.M.; Widdowson, M.A.; et al. Timing of influenza A(H5N1) in poultry and humans and seasonal influenza activity worldwide, 2004–2013. *Emerg. Infect. Dis.* **2015**, *21*, 202–208. [CrossRef] [PubMed]
122. Subbarao, K. Advances in Influenza Virus Research: A Personal Perspective. *Viruses* **2018**, *10*, 724. [CrossRef] [PubMed]

veterinary sciences

Article

Identification and Characterization of Viral and Bacterial Pathogens in Free-Living Bats of Kopaonik National Park, Serbia

Dejan Vidanović [1,*], Nikola Vasković [1], Marko Dmitrić [1], Bojana Tešović [1], Mihailo Debeljak [1], Milovan Stojanović [1] and Ivana Budinski [2]

1 Veterinary Specialized Institute Kraljevo, 36000 Kraljevo, Serbia
2 Department of Genetic Research, Institute for Biological Research "Siniša Stanković"—National Institute of the Republic of Serbia, University of Belgrade, 11108 Belgrade, Serbia
* Correspondence: vidanovic@yahoo.com

Simple Summary: Globalization, climate change, and increasing human-driven environmental changes have elevated the risk of zoonotic disease emergence—diseases that can be transmitted from animals to humans. Bats, while playing vital ecological roles, are recognized as natural reservoirs of numerous pathogens, including over 200 viruses, such as lyssaviruses, filoviruses, coronaviruses, and henipaviruses, some of which are linked to serious diseases like rabies, Ebola, and SARS-CoV. Although viral diversity in bats is well-documented, bacterial pathogens remain underexplored, despite growing evidence of their presence. This pilot study investigated 40 individuals from 12 bat species in Serbia's Kopaonik National Park using microbiological and molecular methods. While no high-risk pathogens, such as SARS-CoV-2, lyssaviruses, or *Salmonella* spp., were found, alphacoronavirus genomes were confirmed in four bats. Additionally, genomes of *Mycoplasma* spp. were present in 45% and *Rickettsia* spp. in 12.5% of individuals, although species-level identification of these pathogens was not possible. These findings highlight the need for continued surveillance of bat-associated microorganisms, particularly in areas where human–wildlife interactions are increasing. Understanding the microbial diversity in bat populations is essential for anticipating potential zoonotic threats and informing public health strategies.

Abstract: This pilot study investigated the presence of potentially zoonotic microorganisms in bat species from Kopaonik National Park, Serbia. A total of 40 individuals from 12 bat species were sampled and screened using microbiological and molecular methods. *Salmonella* spp., *Chlamydia* spp., *Coxiella burnetii*, *Francisella tularensis*, *Leptospira* spp., *Lyssavirus*, Filoviridae, henipaviruses, and SARS-CoV-2 were not detected in any bats. Coronavirus genomes were confirmed in four bats—one *Myotis brandtii*, two *Myotis daubentonii*, and one *Myotis* cf. *mystacinus*. Sequence analysis identified the presence of alphacoronavirus genomes with high similarity to strains previously found in Europe. *Mycoplasma* spp. genomes were found in 18 bats (45%), and *Rickettsia* spp. were detected in five bats (12.5%), although species-level identification was not possible. The findings highlight the presence of certain bacteria and viruses in bats that could have implications for public health, especially in areas with close human–wildlife interaction. Although no direct evidence of high-risk pathogens was found, the results support the importance of continued surveillance and ecological studies on bats, given their role as potential reservoirs. Monitoring bat-associated microorganisms is essential to better understand possible transmission routes and improve the prevention of emerging zoonotic diseases.

https://doi.org/10.3390/vetsci12050401

Keywords: chiroptera; coronavirus; *Mycoplasma*; *Rickettsia*; viruses; bacteria; zoonoses

1. Introduction

Globalization and environmental, climate, and human-induced changes provide ample opportunities for the spread and emergence of zoonoses—diseases that can be transmitted between animals and humans. Today, it is known that about 75% of all infectious diseases in humans are zoonoses, resulting from the transmission of pathogens from domestic or wild animals to humans [1].

Bats (order Chiroptera) are the second most numerous group of mammals in the world, right after rodents. More than 1480 species have been described so far [2], and bats account for more than 20% of all mammalian species. They play an important role in ecosystems worldwide, as suppressors of pest insect populations, seed dispersers, pollinators, etc. However, alongside these ecological benefits, bats have also been increasingly recognized for their role in the transmission of infectious diseases. Growing evidence suggests that bats are significantly underestimated as natural reservoirs of a wide range of viruses, including many of public health significance [3]. Nevertheless, they have a similar proportion of viruses per host species as other orders of mammals [4,5].

To date, more than 200 viruses have been identified in bats, including species from both suborders of Chiroptera: Yinpterochiroptera (including the superfamily Rhinolophoidea and family Pteropodidae) and Yangochiroptera (which includes all other bat families). These viruses belong to 27 different viral families, indicating an extraordinary diversity in bats of virus species [6], some of which are highly pathogenic to humans. Adding to the need for precaution is that bats serve as hosts for the recombination, transmission, and spread of viral pathogens.

Perhaps the best-known example of viruses associated with bats is the rabies virus (family *Rhabdoviridae*), with a link between bats and human infections recognized for over a century—since the first identification of rabies in asymptomatic vampire bats [7]. Other important groups of bat-associated viruses include henipaviruses (Nipah virus and Hendra virus), filoviruses (Marburg virus, Ebola virus, and Mengla virus), bunyaviruses (e.g., Ahun virus), and coronaviruses, such as SARS-CoV and MERS-CoV. SARS-CoV emerged in 2002–2003 in Guangdong Province, China, and bats were later identified as the natural reservoir of this novel pathogen. A decade later, in 2012, MERS-CoV emerged in the Middle East. Although dromedary camels were the direct source of human infections, viruses closely related to MERS-CoV have also been detected in bats, suggesting a possible ancestral origin in bat populations [8].

While global reviews have highlighted the roles of bats as carriers and potential reservoir hosts of human-pathogenic viruses, much less is known about the public health relevance of viruses found specifically in European bat populations. European bats have been increasingly recognized as hosts of a wide diversity of viruses, some of which have zoonotic potential. A comprehensive review by Kohl et al. [9] identified viruses from 21 different viral families in European bats, with *Adenoviridae*, *Rhabdoviridae*, and *Coronaviridae* being the most frequently detected and most intensively studied. Among these, lyssaviruses—particularly European Bat *Lyssavirus* 1 (EBLV-1) and European Bat *Lyssavirus* 2 (EBLV-2)—remain the only bat viruses in Europe with clearly established zoonotic potential, capable of causing fatal rabies-like disease in humans, although spillover events are rare. Molecular surveillance has also revealed the presence of various coronaviruses (mainly alpha- and betacoronaviruses) and novel paramyxoviruses, but none have been linked to human disease so far [9].

In contrast to viruses, bacterial pathogens in bats have been understudied despite their ubiquity and diversity. However, in recent years, this trend has been changing, with an increasing number of studies investigating bacteria (from vector-borne to enteric pathogens), protozoa, parasites, and fungal agents present in bats. The most prevalent pathogens are from the genera *Leptospira*, *Mycoplasma*, *Brucella*, *Coxiella*, *Francisella*, and *Rickettsia* [10]. The dynamics of infection in bats is driven by a complex interaction of environmental, immunological, behavioural, and anthropogenic factors [11]. In the future, it will be crucial to conduct interdisciplinary studies to better understand the circumstances that contribute to the emergence of disease in bat populations and to reduce threats to humans and animals.

To date, 31 species of bats have been recorded in Serbia, indicating the great diversity of this group of mammals in the country [12]. While in some parts of Serbia, bat surveys have been systematically conducted over many years, data on bat fauna are sparse in other areas (including many parts of Central Serbia), with little or no information available on the pathogens associated with bats in Serbia. Remarkably, there is a lack of published studies on bat pathogen investigations within the country.

Kopaonik National Park covers the highest and most well-preserved parts of the Kopaonik mountain, which rises in the central part of southern Serbia. The park covers an area of 11,810 hectares and begins at an altitude of about 800 m and ends with the highest peak 2017 m above sea level. Due to its natural beauty and high level of biodiversity, Kopaonik National Park is a popular tourist attraction, so it could be a hotspot of interaction between bats and humans.

The aim of this pilot study was to detect and classify pathogenic microorganisms in bat species inhabiting the Kopaonik National Park area in order to improve knowledge about the pathogens' distribution and to create control and surveillance measures for their potential transmission to humans.

2. Materials and Methods

2.1. Study Area and Sample Collection

During the summer and autumn of 2024 (8–9 July 2024, 21–22 August 2024, and 10–11 October 2024), three field studies were conducted in Kopaonik National Park, Serbia. Bats were trapped at several sites at the locations Metođe (geo: 43.2694, 20.8247, altitude 1540 m) and Zaplanina (geo: 43.2651, 20.8598, altitude 1104 m). Before trapping, acoustic monitoring across various sites within the park, covering different habitats and altitudes, was conducted. Based on the areas with the highest bat activity and species diversity revealed through acoustic surveys, specific trapping sites were selected. Trapping was carried out at the entrances of mine shafts (where a higher number of bat passes were recorded) as well as along forest trails where bat activity was also detected. For trapping, standard vertical nets (mist-nets, Ecotone, Poland) 3–12 m in length were set up near entrances to abandoned mine shafts and across forest trails. Nets were opened at sunset and taken down 4 h later. Trapped bats were identified to the species level following Dietz and Kiefer [13], sexed, and their age and reproductive status were assessed. Four swabs (two oral and two anal) were collected from all individuals for bacteriological and virological analysis. Bats were released immediately after processing and sample collection at the trapping sites. Trapping and sample collection were carried out under permit no. 000493134 2023 14850 004 003 501 086 issued by the Ministry for Environmental Protection of the Republic of Serbia.

2.2. Bat Pathogen Analyses

Analyses of pathogens were carried out at the Veterinary Specialized Institute Kraljevo. The examined pathogens were as follows:

(1) Viruses: lyssaviruses (family *Rhabdoviridae*); henipaviruses (Nipah virus and Hendra virus); Lloviu virus (family *Filoviridae*); alpha-, beta-, gamma-, and deltacoronaviruses; and SARS-CoV-2 (family *Coronaviridae*).

(2) Bacteria: Salmonella, Leptospira, Mycoplasma, Coxiella, Brucella, and Rickettsia.

Swabs were kept on dry ice until they were brought to the laboratory, where they were stored at −80 °C until the beginning of the examination. For each bat, one oral and one anal swab were pooled and used for *Salmonella* isolation, while the second oral and anal swabs were pooled and then used for DNA/RNA extraction and molecular investigation.

Salmonella was isolated following the accredited method EN ISO 6579 (ISO 6579-1:2017, "Microbiology of the food chain—Horizontal method for the detection, enumeration, and serotyping of *Salmonella*") [14]. Briefly, swabs were first cultured in pre-enrichment buffered peptone water (BPW) for 18 ± 3 h at 37 °C. Incubated pre-enrichment medium (0.1 mL) was transferred to modified semi-solid Rappaport–Vassiliadis (MSRV) agar and incubated at 40.5–42.5 °C. After 24 h incubation of selective enrichment, bacterial growth from the edge of the growth zone was streaked onto plates, one containing xylose–lysine–deoxycholate (XLD) agar and one brilliant green agar (BGA).

For genomic DNA and RNA extraction, oral and rectal swabs from each bat were placed together in 1.5 mL tubes with the addition of 800 µL of phosphate-buffered saline. After vortexing and brief centrifugation, 200 µL of supernatant was transferred to a fresh tube and extracted using a Bioextract Superball extraction kit (Biosellal, Dardilly, France) following the manufacturer's instructions. A Luna universal probe One Step RT-PCR kit (NEB, Ipswich, MA, USA) was used for real-time and RT-PCR reactions, Luna universal RT-PCR mastermix (NEB, USA) was used for SYBR RT-PCR reactions, and Luna universal probe qPCR master mix was used for real-time PCR reactions. Real-time PCR reactions were performed in an AriaMx thermal cycler (Agilent, Penang, Malaysia), and RT-PCR reactions were performed in a 2720 thermal cycler (Applied Biosystems, Singapore).

The presence of lyssaviruses was determined by using the Pan-lyssavirus Real Time RT-PCR assay for Rabies Diagnosis [15]. Filoviruses were determined using two assays, RT-PCR described by He et al. [16] and SYBR RT-PCR described by Coertse et al. [17]. Henipaviruses and Nipah and Hendra viruses were determined using RT-PCR assays described in [18]. Coronaviruses were determined using two assays, i.e., a nested RT-PCR assay [19] and a RT-PCR assay [20], and SARS-CoV-2 virus was determined using the real-time RT-PCR assay developed by Corman et al. [21]. Mycoplasmas were determined using the PCR assay described by Cultrera et al. [22]; Chlamydiaceae were determined using the real-time PCR assay described by Ehricht et al. [23]; *Coxiella burnetii* was determined using the real-time PCR assay described by Klee et al. [24]; and Rickettsiae were determined using the real-time PCR assay described by Kato et al. [25]. Pathogenic *Leptospira* were determined using the Real Time PCR assay described by Stoddard et al. [26]. Primers and probes used in this study are listed in Table 1.

PCR products of coronaviruses and mycoplasmas were purified and sequenced with a Sanger SeqStudio sequencer (ThermoFisher Scientific, Waltham, MA, USA) externally. The obtained sequences were processed in Chromas lite 2.1 software and compared with other coronavirus sequences in National Centre for Biotechnology Information (NCBI, Betesda, ML, USA) database using free online Basic Local Alignment Search Tool (BLAST) software. Mega 12 software was used for phylogenetic analysis.

Table 1. Primers and probes used in this study.

Assay	Primer/Probename	Primer/Probe Sequences
Lyssavirus [15]	JW12	ATG TAA CAC CYC TAC AAT G
	N165	GCA GGG TAY TTR TAC TCA TA
Filoviruses [16]	Filo-F	TGATATATGATCATCTTCCAGG
	Filo-in-F	GCATTTCACCAATTAACACAGG
	Filo-R	TTTATATGAATCAGTGGAGGTG
	FV-F1	GCMTTYCCIAGYAAYATGATGG
	FV-R1	GTDATRCAYTGRTTRTCHCCCAT
	FV-F2	TDCAYCARGCITCDTGGCAYC
	FV-R2	GIGCACADGADATRCWIGTCC
Filoviruses [17]	Filo-SYBR-F	GRGARTAYGCICCITTYGC
	Filo-SYBR-R	AGYTGYTGRTAYTGYTCICC
Hendra [18]	HeV M 5755F	CTTCGACAAAGACGGAACCAA
	HeV M 5823R	CCAGCTCGTCGGACAAAATT
Nipah [18]	NiV_N_1198F	TCAGCAGGAAGGCAAGAGAGTAA
	NiV_N_1297R	CCCCTTCATCGATATCTTGATCA
Pan-corona [19]	Pan_CoV_F1	GGTTGGGAYTAYCCHAARTGYGA
	Pan_CoV_R1	CCRTCATCAGAHARWATCAT
	Pan_CoV_R2	CCRTCATCACTHARWATCAT
	Pan_CoV_F2	GAYTAYCCHAARTGTGAYAGA
	Pan_CoV_F3	GAYTAYCCHAARTGTGAYMGH
Coronavirus [20]	Watanabe conventional_F	GGTTGGGACTATCCTAAGTGTGA
	Watanabe conventional_R	CCATCATCAGATAGAATCATCATA
SARS-CoV2 [21]	SARS2-IP4-14059F	GGT AAC TGG TAT GAT TTC G
	SARS2-IP4-14146R	CTG GTC AAG GTT AAT ATA GG
	SARS2-IP4-14084FAM	FAM-TCA TAC AAA CCA CGC CAG G-BHQ1
Mycoplasma [22]	GPO-1	ACTCCTACGGGAGGCAGCAGTA
	MGSO	TGCACCATCTGTCACTCTGTTAACCTC
Chlamydiae [23]	Ch23S-F	CTGAAACCAGTAGCTTATAAGCGGT
	Ch23S-R	ACCTCGCCGTTTAACTTAACTCC
	Ch23S-Pro	FAM-CTCATCATGCAAAAGGCACGCCG-BHQ1
Coxiella burnetii [24]	Cox-F	GTCTTAAGGTGGGCTGCGTG
	Cox-R	CCCCGAATCTCATTGATCAGC
	Cox-TM	FAM-AGCGAACCATTGGTATCGGACGTT-TAMRA-TATGG
Rickettsia [25]	PanR8_F	AGCTTGCTTTTGGATCATTTGG
	PanR8_R	TTCCTTGCCTTTTCATACATCTAGT
	PanR8_P	FAM-CCTGCTTCTATTTGTCTTGCAGTAACACGCCA-BHQ1
Pathogenic *Leptospira* spp. [26]	LipL32-45F	AAGCATTACCGCTTGTGGTG
	LipL32-286R	GAACTCCCATTTCAGCGATT
	LipL32-189P	FAM-AAAGCCAGGACAAGCGCCG-BHQ1

3. Results

A total of 40 bats from 12 different species were captured and sampled: *Rhinolophus ferrumequinum*, *Rhinolophus hipposideros*, *Barbastella barbastellus*, *Myotis alcathoe*, *Myotis blythii*, *Myotis brandtii*, *Myotis cf. mystacinus*, *Myotis daubentonii*, *Myotis emarginatus*, *Myotis myotis*, *Myotis nattereri*, and *Plecotus auritus*.

Salmonella was not isolated by the culture method from any of the bats. The genomes of lyssaviruses, filoviruses, henipaviruses (Hendra and Nipah), and SARS-CoV-2 viruses were not detected using molecular methods. Coronavirus genomes were detected in 4 of 40 bats,

1 in the bat species *Myotis brandtii* (No. 4), 2 in *Myotis daubentonii* bats (No. 6 and No. 8), and 1 in a *Myotis* cf. *mystacinus* bat (No. 35) (Table 2). Coronaviruses in bats No. 4 and No. 6 were detected using the Holbrook nested RT-PCR assay [19], and coronaviruses in bats No. 8 and No. 35 were detected using Watanabe RT-PCR assay [20]. Sequencing of part of the coronavirus genome from bat No. 4 (271 nucleotides) classified it as a bat alphacoronavirus, with 98.56% similarity to a virus from Russia (Bat-CoV/M.brandtii/Russia/MOW15-27/1, OQ725983.1) detected in *M. brandtii*. The same was true for the genome from bat No. 6 (337 nucleotides), which also showed 98.56% similarity to the same virus from Russia. The genomes of the viruses from bats No. 4 and No. 6 differed by only one nucleotide (99.6% similarity). Bat coronavirus from bat No. 8 had 99.20% similarity with a bat coronavirus isolate from Denmark, BtCoV/13585-58/M.dau/DK/2014 (MN535732.1), detected in *M. daubentonii*, while bat coronavirus from bat No. 35 had 93.92% similarity with a bat coronavirus isolated from Russia, BtCoV/631/RUS/2020 (MZ322309.1), detected in *M. daubentonii*.

Table 2. Pathogen genome presence in bats from Kopaonik National Park, Serbia.

Bat No.	Bat Species	Coronaviruses	*Mycoplasma* sp.	*Rickettsia* sp.
1	*Rhinolophus ferrumequinum*	−	+	−
2	*Rhinolophus ferrumequinum*	−	+	−
3	*Rhinolophus hipposideros*	−	−	−
4	*Myotis brandtii*	+	+	+
5	*Myotis brandtii*	−	−	−
6	*Myotis daubentonii*	+	+	−
7	*Myotis* cf. *mystacinus*	−	−	+
8	*Myotis daubentonii*	+	−	−
9	*Barbastella barbastellus*	−	−	+
10	*Myotis blythii*	−	+	−
11	*Myotis myotis*	−	+	−
12	*Rhinolophus ferrumequinum*	−	+	−
13	*Myotis daubentonii*	−	−	−
14	*Myotis nattereri*	−	−	−
15	*Myotis alcathoe*	−	−	−
16	*Myotis* cf. *mystacinus*	−	−	−
17	*Myotis brandtii*	−	−	+
18	*Myotis daubentonii*	−	−	−
19	*Rhinolophus ferrumequinum*	−	+	−
20	*Myotis myotis*	−	+	−
21	*Plecotus auritus*	−	−	+
22	*Plecotus auritus*	−	−	−
23	*Plecotus auritus*	−	−	−
24	*Myotis emarginatus*	−	+	−
25	*Myotis emarginatus*	−	+	−
26	*Myotis daubentonii*	−	−	−
27	*Myotis* cf. *mystacinus*	−	+	−
28	*Myotis daubentonii*	−	−	−
29	*Plecotus auritus*	−	+	−
30	*Myotis myotis*	−	−	−
31	*Rhinolophus ferrumequinum*	−	+	−
32	*Plecotus auritus*	−	+	−
33	*Plecotus auritus*	−	+	−
34	*Rhinolophus ferrumequinum*	−	+	−
35	*Myotis* cf. *mystacinus*	+	−	−
36	*Myotis myotis*	−	+	−
37	*Plecotus auritus*	−	−	−
38	*Barbastella barbastellus*	−	−	−
39	*Myotis daubentonii*	−	−	−
40	*Rhinolophus hipposideros*	−	−	−

Phylogenetic analysis (Figure 1) of part of the RdRp gene showed that the alphacoronaviruses detected in bats No. 8 and No. 35 were classified in the subgenus *Pedacovirus*, while the viruses detected in bats No. 4 and No. 6 were classified in the subgenus *Myotacovirus*.

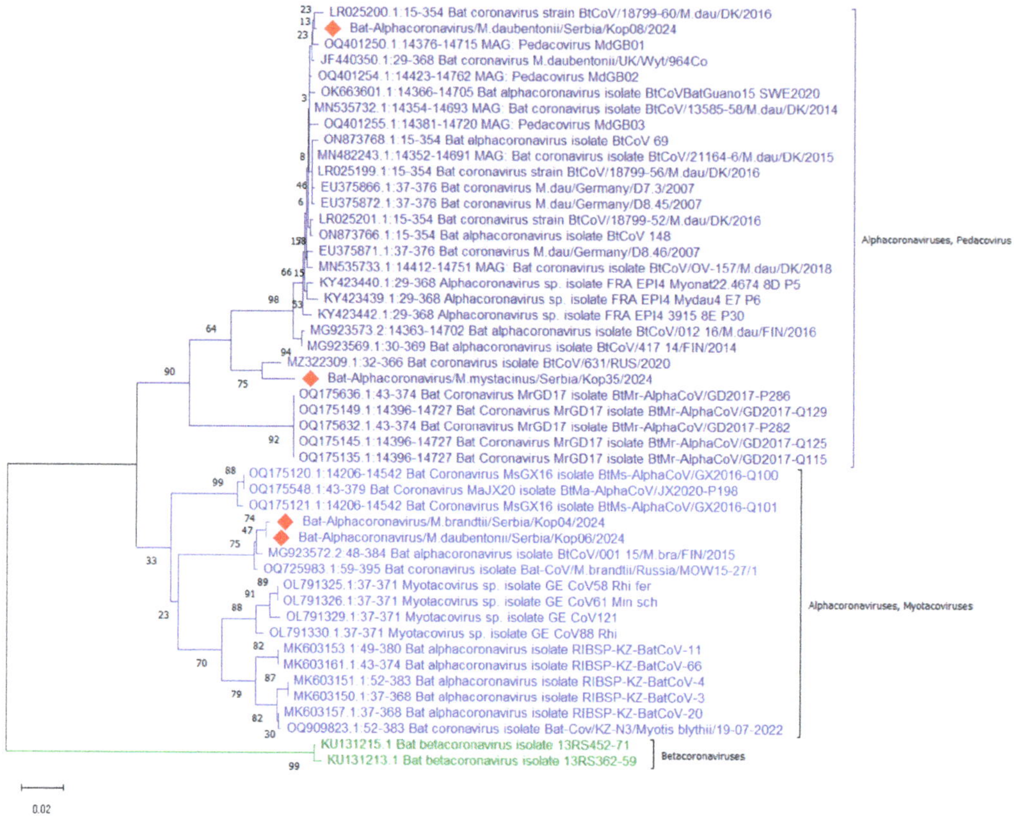

Figure 1. Evolutionary relationships of 48 taxa based on partial RdRp gene.

The evolutionary history was inferred using the Neighbor-Joining method. The optimal tree with the sum of branch length = 0.656 is shown. The percentages of replicate trees in which the associated taxa clustered together in the bootstrap test (1000 replicates) are shown next to the branches. The tree is drawn to scale, with branch lengths in the same units as those of the evolutionary distances used to infer the phylogenetic tree. The evolutionary distances were computed using the Maximum Composite Likelihood method and are in units of the number of base substitutions per site. The analytical procedure encompassed 48 coding nucleotide sequences using the first, second, third, and non-coding positions. The pairwise deletion option was applied to all ambiguous positions for each sequence pair, resulting in a final data set comprising 304 positions. Evolutionary analyses were conducted in MEGA12 [27]. Sequences of alphacoronaviruses from Serbia are labelled with red diamond.

Mycoplasma genomes were detected in 18 of 40 bats (45%), the species being *R. ferrumequinum*, *M. brandtii*, *M. daubentonii*, *M. blythii*, *M. myotis*, *M. emarginatus*, *M.* cf. *mystacinus*, and *P. auritus*. Nucleotide sequencing of PCR products could not reveal the species of *Mycoplasma* but only the genera. *Rickettsia* species genomes were detected in 5 of 40 bats (12.5%), with the species being *M. brandtii*, *M.* cf. *mystacinus*, *B. barbastellus*, and *P. auritus*.

The genomes of other pathogenic bacteria, i.e., *Chlamydia* spp., *Coxiella burnetii*, pathogenic *Leptospira* spp., and *Francisella tularensis*, were not detected in the examined bats. An overview of the microorganism genomes confirmed in the analysed bats is given in Table 2.

4. Discussion

Coronaviruses

Coronaviruses (order Nidovirales, family *Coronaviridae*) include four genera: alphacoronavirus and betacoronavirus, which infect a wide range of mammals, and gammacoronavirus and deltacoronavirus, which primarily infect birds. Coronaviruses have been identified in 238 species of bats worldwide. Alphacoronaviruses and betacoronaviruses have been identified in bats from 14 of the 21 bat families, in at least 69 countries on six continents [28]. In Europe, coronaviruses have been recorded in the bat families Rhinolophidae, Vespertilionidae, and Miniopteridae, in nine genera (*Rhinolophus, Myotis, Pipistrellus, Plecotus, Tadarida, Nyctalus, Hypsugo, Eptesicus,* and *Miniopterus)* and, in total, in 26 species [29], including *Myotis brandtii* and *Myotis daubentonii*. Phylogenetic analysis of alphacoronaviruses detected in Kopaonik National Park classified them as belonging to two subgenera of alphacoronaviruses, *Myotacovirus*, detected in *M. brandtii* and *M. daubentonii*, and *Pedacovirus*, detected in *M. daubentonii* and *M.* cf. *mystacinus*. Our literature review did not reveal any previous studies confirming the presence of alphacoronaviruses in *M. mystacinus* bats. These results are in accordance with the results published previously by researchers from the Scientific Institute of Veterinary Medicine Novi Sad, Serbia, who determined the presence of alpha and betacoronaviruses in 35 out of 142 samples in the territory of Vojvodina province (Serbia) [30].

Alphacoronaviruses mainly infect bats, but they have the potential to transfer to humans or other animals and cause disease. Although these viruses have not yet caused significant outbreaks among humans, they are considered a potential organism for future zoonotic transmission [31], particularly in areas where there is close contact between humans and bats (e.g., caves) or habitats with a high abundance of bats. Recent research shows that all currently known human coronaviruses have an animal origin, with bat or rodent coronaviruses being the most probable ancestors. In most cases, it was suggested that other mammals served as intermediate hosts prior to final adaptation to humans [32].

Analysis of the genome sequence identity and geographic distribution of the homologous alphacoronaviruses examined in this pilot study in Kopaonik National Park provides valuable information into the broad geographic distribution and evolutionary dynamics of these viruses. Alphacoronaviruses from bats in Kopaonik National Park had high sequence identity (up to 99%) with bat alphacoronaviruses from Russia, Denmark, Finland, and Germany; this is in agreement with results obtained by other researchers [33,34]. This extensive geographic distribution highlights the ecological adaptability of alphacoronaviruses and their potential to spread across large and diverse regions, most likely via the short- and long-distance movements of bats [34]. Bats are capable of flight, enabling them to maintain and spread the pathogens they carry. These pathogens include vector-borne parasites and bacteria, which can also be transmitted through ectoparasites such as bat flies and ticks [35].

The investigation and monitoring of coronaviruses is crucial in order to detect in time possible mutations that could increase their transmissibility or pathogenicity from bats to humans.

Lyssaviruses, Filoviruses

Neither lyssavirus nor filovirus genomes were detected in the 40 bats from Kopaonik National Park. The results of this study are in accordance with earlier research on lyssaviruses in bats in Serbia, where these viruses were not detected. The only reported case of lyssavirus in bats in Serbia was in 1954 [36], but since then, there have been no reported cases. In 2006, a two-year study of bats did not reveal any positive case [37]. Elsewhere in Europe, lyssaviruses have been detected in several countries, including the United Kingdom, the Netherlands, Finland, Germany, Spain, Italy, Slovenia, Croatia, Hungary, Bulgaria, Ukraine, and Russia. As expected, filovirus genomes, such as Ebola virus and Marburg virus, were not found in any bat from Kopaonik National Park, as these viruses are associated with bats from the family Pteropodidae that is not present in Serbia. Lloviu virus genome was also not detected in the Kopaonik bats, although it was previously detected in *Miniopterus schreibersii* bats in other locations in Serbia, neighbouring Bosnia and Herzegovina, and Hungary [38]. *M. schreibersii* is the main host of this filovirus [39], but during the current research, no bats of this species were captured in Kopaonik National Park.

Mycoplasma spp.

The occurrence of mycoplasmas in bats has potential ecological and epidemiological significance for both human and domestic animal health. Mycoplasmas are distributed worldwide and are present in various species of bats, although a higher prevalence of mycoplasmas has been observed in vampire bats (Desmodontinae) compared to other species [40]. Some *Mycoplasma* species are zoonotic, meaning they can be transmitted from bats to humans or domestic animals. Given that mycoplasmas frequently colonize the respiratory and urogenital tracts, their role in causing disease in domestic animals, such as cattle, pigs, horses, or poultry, is possible. Mycoplasmas have already been recognized as causative agents of diseases such as swine enzootic pneumonia (*Mycoplasma hyopneumoniae*) or bovine infectious pneumonia (*Mycoplasma bovis*). Contact between bats and domestic animals, especially in rural areas, increases the risk of transmission. In some studies, the prevalence of *Mycoplasma* in bats ranged from 3.2% [41] up to 47% [42]. Although mycoplasmas generally cause mild or opportunistic infections in humans, certain species, such as *Mycoplasma pneumoniae*, the cause of atypical pneumonia, can cause more serious problems [35].

Among the 40 bats from Kopaonik National Park studied, 18 (45%) harboured mycoplasma genomes: *Rhinolophus ferrumequinum* (6×), *Plecotus auritus* (3×), *Myotis myotis* (3×), *Myotis emarginatus* (2×), *Myotis blythii* (1×), *Myotis brandtii* (1×), *Myotis* cf. *mystacinus* (1×), *Myotis daubentonii* (1×). Findings of mycoplasmas and their genomes in bats indicate their potential importance as reservoirs for these bacteria, requiring further research to understand the risks to humans and domestic animals. Monitoring and studying these microorganisms can contribute to better prevention of zoonotic infections and preservation of human and animal health.

Rickettsia spp.

Bacteria of the genus *Rickettsia* are intracellular pathogens transmitted by vectors, such as ticks, fleas, and lice, and are known to cause a variety of diseases in humans and animals, including rickettsioses (such as typhus and spotted fever). The presence of *Rickettsia* in bats is of potential importance to human and domestic animal health due to the possibility of zoonotic transmission, although this area is not yet sufficiently investigated. If bats serve as reservoirs for *Rickettsia* bacteria, there is a potential risk that the bacteria could be transmitted to humans by ectoparasites living on bats. Given that bats are common in urban and rural areas, contact with their habitats or ectoparasites may increase the possibility of outbreaks of zoonotic diseases. The presence of *Rickettsia* in bats could also pose a risk to domestic and wild animals, especially those that share a habitat with bats or are in contact with the same vectors. There are currently not many data on the prevalence and species of *Rickettsia* in bats, and even though these potentially zoonotic bacterial pathogens are frequently detected in bat-associated ectoparasites, their role in pathogen transmission remains poorly understood [40]. *Rickettsia* genomes were detected in five of 40 bats (12.5%) from Kopaonik National Park: *Plecotus auritus* (1×), *Myotis brandtii* (2×), *Myotis* cf. *mystacinus* (1×), and *Barbastella barbastellus* (1×). Our findings are in agreement with the findings of researchers from Slovenia who reported the presence of rickettsiae in 17% of bats [43] but, unlike their study, where the highest prevalence of *Rickettsia* was found in the genus *Rhinolophus*, our study did not detect *Rickettsia* in this genus but rather in the genus *Myotis*. Since the aforementioned study employed NGS metagenomic analysis of faecal samples, the authors concluded that caution should be exercised when interpreting the results, as identification was achieved at the *Rickettsia* genus level rather than the species level. They recommended species-specific PCR for further identification of *Rickettsia* [41]. A similar study was conducted in Romania [44], where tissue samples from dead bats were used. Species-specific PCR was performed and 17 sequenced samples had 99.7–100% genetic identity to *Rickettsia monacensis* strains from *Ixodes ricinus* ticks in Romania, Italy, and Serbia. Of the 17 sequenced samples, 5 were from *Pipistrellus pipistrellus* and 12 from *Nyctalus noctula*. That study [44] reported the first detection of *R. monacensis* in European bat tissues and the first global record in these two bat species. While Spotted Fever Group (SFG) rickettsiae have previously been found in ectoparasites associated with bats, their presence in bat tissues suggests that bats may play a role in the transmission or maintenance of *Rickettsia* [44]. The presence of *Rickettsia* genomes in bats in Kopaonik National Park highlights the need for further research. The first step would be to sequence the genomes obtained in order to identify the exact species, followed by attempts to determine the role of bats as a reservoir and the possibility of zoonotic transmission of *Rickettsia*. A better understanding of transmission dynamics, as well as vector identification, is essential for risk assessment and the development of preventive measures. This includes monitoring bat habitats, vector control, and education about potential risks in communities that come into close contact with bats.

Other potentially pathogenic bacteria in bats

Salmonella was not isolated from the 40 bats from Kopaonik National Park. However, several members of the Enterobacteriaceae family, including *Salmonella*, have previously been isolated from bats. Studies have shown that *Salmonella* serotypes isolated from bats have similar characteristics to those found in domestic animals and humans, suggesting that bats may be locally important in the epidemiology of salmonellosis in humans and domestic animals [45]. Some of the serotypes isolated from bats, i.e., *S. Typhimurium*, *S. Enteritidis*, and *S. Virchow*, are common causes of disease in humans and animals [46].

Coxiella, *Leptospira*, and *Chlamydia* genomes were not detected in the 40 Kopaonik bats. *Coxiella burnetti* has previously been reported in bats, but the role of these animals in the spread of *Coxiella* has not yet been fully investigated [47]. Bats have recently been identified worldwide as an important reservoir of various *Leptospira* species (*L. interrogans*, *L. borgpetersenii*, *L. kirschneri*, *L. fainei*, and *L. noguchii*). Currently, more than 107 species of bats have been reported to harbour these bacteria [48], but as prevalences range from 0% to 85.7% [40], the absence of signals from *Leptospira* in the Kopaonik bats is unsurprising. Chlamydiae have not been studied in bats in Serbia to date, so this report on the absence of their signals in the Kopaonik bats is new to the scientific record.

This study presents the first results of microbiological research on bats in Kopaonik National Park, contributing to a broader understanding of the presence of pathogenic microorganisms in the region. However, some limitations of this study must be considered. The sample size was relatively small, primarily due to the high altitude of the National Park. Previous knowledge about bat distribution within Kopaonik National Park was scarce prior to this research, and no data on known bat roosts were available. To survey bat presence, potential roost sites (such as abandoned buildings and mine shafts) were searched, since there are no natural caves in the area. Given the high altitude of the study area, nursery colonies were not expected to be found, even though sampling was during the summer and autumn (females of some bat species form large nursery colonies during the reproductive period). This assumption was confirmed, as only adult individuals were trapped during the summer and autumn of 2024. It is likely that bats use the mine shafts as transitional and/or swarming roosts. There are some bat records from Serbia at the altitude of Kopaonik, but very few bat roosts are known at this elevation [12]. Therefore, since the presence of numerous colonies was not anticipated, we did not expect to capture a large number of individuals. However, despite the relatively low number of trapped bats ($n = 40$), species diversity was greater than expected ($n = 12$). This large species diversity, revealed for the first time in our pilot study, should stimulate further much-needed research into the bat populations in Kopaonik National Park.

Live animals were captured and not euthanized in our pilot study. No carcasses of dead bats were found during the sampling from which internal organs could have been collected, potentially containing pathogens that are not present in oral or rectal swabs. PCR-based screening, although highly sensitive, allows only known or closely related pathogens to be detected, potentially failing to capture more divergent or novel viruses.

This pilot study focused on targeted pathogenic microorganisms previously documented in the literature as animal or zoonotic pathogens. However, confirming and expanding these findings requires a larger number of samplings throughout the year to obtain a larger sample size, including ectoparasites such as bat flies and ticks. Future research should also include next-generation sequencing (NGS) metagenomic analysis to comprehensively characterize the microbial communities present in bats. Metagenomic approaches would enable unbiased detection of both known and novel pathogens.

5. Conclusions

Bats are of great importance in the preservation of the entire ecosystem, but at the same time, they can be reservoirs of various pathogens, including viruses and bacteria with zoonotic potential, which pose a potential threat to human and animal health. Bats from Kopaonik National Park harboured the genomes of some potentially zoonotic microorganisms, such as alphacoronaviruses, mycoplasmas, and rickettsiae, while genomes of other, more dangerous pathogens, such as coronaviruses related to SARS-CoV-2, lyssaviruses, filoviruses, and paramyxoviruses, were not found in the animals.

The results of the virological tests were expected, because so far, lyssaviruses and paramyxoviruses have not been detected in Serbia, primarily due to the absence of bat species that are their carriers. The finding of alphacoronaviruses were also expected, given their well-documented presence in Serbian and European bat populations.

Although this pilot study was conducted on a small number of bats from a relatively small geographical area, it is the first systematic pathogen surveillance in bats conducted in Kopaonik National Park and provides a good starting point for further research. The results highlight the need for continuous monitoring and longitudinal study of bats as reservoirs of pathogens. Implementing regular monitoring programs would significantly improve the country's capacity for early detection of pathogens, which is crucial for improving the prevention of animal diseases and zoonoses and for supporting One Health initiatives aimed at protecting human and animal health.

Author Contributions: Conceptualization, D.V. and I.B.; methodology, D.V., M.D. (Marko Dmitrić) and B.T.; software, M.S.; validation, D.V. and I.B.; formal analysis, D.V., B.T.; investigation, M.D. (Mihailo Debeljak) and N.V.; resources, D.V. and I.B.; data curation, M.D. (Marko Dmitrić); writing—original draft preparation, D.V.; writing—review and editing, D.V. and I.B.; visualization, B.T.;

supervision, D.V.; project administration, D.V.; funding acquisition, N.V. All authors have read and agreed to the published version of the manuscript.

Funding: This research was funded by Kopaonik National Park, Serbia, contract numbers 665/24 and 2-163.

Institutional Review Board Statement: Bat trapping and sample collection were carried out under permit no. 000493134 2023 14850 004 003 501 086 issued by the Ministry for Environmental protection of Republic of Serbia.

Informed Consent Statement: Not applicable.

Data Availability Statement: All data are contained within this paper.

Acknowledgments: Authors are grateful for field assistance and institutional support provided by Kopaonik National Park. Bat nets were obtained from the International Atomic Energy Agency (IAEA) through project RER/5/027 and diagnostic procedures supported through project SRB/5/005. IB was supported by the Ministry of Science, Technological Development and Innovation of the Republic of Serbia, contract no. 451-03-136/2025-03/200007. The results presented in this manuscript are in line with Sustainable Development Goal 15 (Life on Land) of the United Nations 2030 Agenda.

Conflicts of Interest: The authors declare no conflicts of interest. The funders had no role in the design of the study; in the collection, analyses, or interpretation of data; in the writing of the manuscript; or in the decision to publish the results.

References

1. Biswas, R.; Debnath, C.; Samanta, I.; Barua, R.; Singh, A.D. Ecology of Bats and Their Role in Emerging Zoonotic Diseases: A Review. *Rev. Sci. Tech. OIE* **2020**, *39*, 1077–1090. [CrossRef] [PubMed]
2. Simmons, N.B.; Cirranello, A.L. Bat Species of the World: A Taxonomic and Geographic Database. Version 1.6. Available online: https://batnames.org/home.html (accessed on 8 December 2024).
3. Fenton, M.B.; Simmons, N.B. *Bats: A World of Science and Mystery*; University of Chicago Press: Chicago, IL, USA, 2015; ISBN 978-0-226-06512-0.
4. Calisher, C.H.; Childs, J.E.; Field, H.E.; Holmes, K.V.; Schountz, T. Bats: Important Reservoir Hosts of Emerging Viruses. *Clin. Microbiol. Rev.* **2006**, *19*, 531–545. [CrossRef]
5. Wang, L.-F.; Walker, P.J.; Poon, L.L.M. Mass Extinctions, Biodiversity and Mitochondrial Function: Are Bats 'Special' as Reservoirs for Emerging Viruses? *Curr. Opin. Virol.* **2011**, *1*, 649–657. [CrossRef]
6. Young, C.C.W.; Olival, K.J. Optimizing Viral Discovery in Bats. *PLoS ONE* **2016**, *11*, e0149237. [CrossRef] [PubMed]
7. Carini, A. Sur une grande epizootie de rage. *Ann. L'inst. Pasteur* **1911**, *25*, 843–846.
8. Farina, L.L.; Lankton, J.S. Chiroptera. In *Pathology of Wildlife and Zoo Animals*; Elsevier: Amsterdam, The Netherlands, 2018; pp. 607–633, ISBN 978-0-12-805306-5.
9. Kohl, C.; Nitsche, A.; Kurth, A. Update on Potentially Zoonotic Viruses of European Bats. *Vaccines* **2021**, *9*, 690. [CrossRef]
10. Chomel, B.B.; Boulouis, H.-J.; Chang, C.; Setién, A.A.; Stuckey, M.J. Bat-Related Zoonoses. In *Zoonoses: Infections Affecting Humans and Animals*; Sing, A., Ed.; Springer International Publishing: Cham, Switzerland, 2022; pp. 1–36, ISBN 978-3-030-85877-3.
11. Hayman, D.T.S.; Bowen, R.A.; Cryan, P.M.; McCracken, G.F.; O'Shea, T.J.; Peel, A.J.; Gilbert, A.; Webb, C.T.; Wood, J.L.N. Ecology of Zoonotic Infectious Diseases in Bats: Current Knowledge and Future Directions. *Zoonoses Public Health* **2013**, *60*, 2–21. [CrossRef]
12. Paunović, M.; Karapandža, B.; Budinski, I.; Stamenković, S. *Bats (Mammalia, Chiroptera) of Serbia*; Serbian Academy of Sciences and Arts: Beograd, Serbia, 2020; Volume 13, ISBN 978-86-7025-887-7.
13. Dietz, C.; Kiefer, A. *Bats of Britain and Europe*; Bloomsbury Publishing: London, UK, 2016; ISBN 978-1-4729-2202-1.
14. *ISO 6579-1:2017*; Microbiology of the Food Chain—Horizontal Method for the Detection, Enumeration, and Serotyping of Salmonella 2017. International Organization for Standardization: Geneva, Switzerland, 2017.
15. Marston, D.A.; Jennings, D.L.; MacLaren, N.C.; Dorey-Robinson, D.; Fooks, A.R.; Banyard, A.C.; McElhinney, L.M. Pan-Lyssavirus Real Time RT-PCR for Rabies Diagnosis. *JoVE* **2019**, *149*, 59709. [CrossRef]
16. He, B.; Feng, Y.; Zhang, H.; Xu, L.; Yang, W.; Zhang, Y.; Li, X.; Tu, C. Filovirus RNA in Fruit Bats, China. *Emerg. Infect. Dis.* **2015**, *21*, 1675–1677. [CrossRef]
17. Coertse, J.; Mortlock, M.; Grobbelaar, A.; Moolla, N.; Markotter, W.; Weyer, J. Development of a Pan-Filoviridae SYBR Green qPCR Assay for Biosurveillance Studies in Bats. *Viruses* **2023**, *15*, 987. [CrossRef]

18. WOAH. Chapter 3.1.16 Nipah and Hendra Virus Diseases (Version Adopted in May 2022). In *Manual of Diagnostic Tests and Vaccines for Terrestrial Animals*; WOAH: Paris, France, 2024.
19. Holbrook, M.G.; Anthony, S.J.; Navarrete-Macias, I.; Bestebroer, T.; Munster, V.J.; Van Doremalen, N. Updated and Validated Pan-Coronavirus PCR Assay to Detect All Coronavirus Genera. *Viruses* **2021**, *13*, 599. [CrossRef] [PubMed]
20. Watanabe, S.; Masangkay, J.S.; Nagata, N.; Morikawa, S.; Mizutani, T.; Fukushi, S.; Alviola, P.; Omatsu, T.; Ueda, N.; Iha, K.; et al. Bat Coronaviruses and Experimental Infection of Bats, the Philippines. *Emerg. Infect. Dis.* **2010**, *16*, 1217–1223. [CrossRef]
21. Corman, V.M.; Landt, O.; Kaiser, M.; Molenkamp, R.; Meijer, A.; Chu, D.K.; Bleicker, T.; Brünink, S.; Schneider, J.; Schmidt, M.L.; et al. Detection of 2019 Novel Coronavirus (2019-nCoV) by Real-Time RT-PCR. *Eurosurveillance* **2020**, *25*, 2000045. [CrossRef] [PubMed]
22. Cultrera, R.; Seraceni, S.; Germani, R.; Contini, C. Molecular Evidence of *Ureaplasma urealyticum* and *Ureaplasma parvum* Colonization in Preterm Infants during Respiratory Distress Syndrome. *BMC Infect. Dis.* **2006**, *6*, 166. [CrossRef] [PubMed]
23. Ehricht, R.; Slickers, P.; Goellner, S.; Hotzel, H.; Sachse, K. Optimized DNA Microarray Assay Allows Detection and Genotyping of Single PCR-Amplifiable Target Copies. *Mol. Cell. Probes* **2006**, *20*, 60–63. [CrossRef]
24. Klee, S.R.; Tyczka, J.; Ellerbrok, H.; Franz, T.; Linke, S.; Baljer, G.; Appel, B. Highly Sensitive Real-Time PCR for Specific Detection and Quantification of *Coxiella burnetii*. *BMC Microbiol.* **2006**, *6*, 2. [CrossRef]
25. Kato, C.Y.; Chung, I.H.; Robinson, L.K.; Austin, A.L.; Dasch, G.A.; Massung, R.F. Assessment of Real-Time PCR Assay for Detection of *Rickettsia* spp. and *Rickettsia rickettsii* in Banked Clinical Samples. *J. Clin. Microbiol.* **2013**, *51*, 314–317. [CrossRef] [PubMed]
26. Stoddard, R.A.; Gee, J.E.; Wilkins, P.P.; McCaustland, K.; Hoffmaster, A.R. Detection of Pathogenic Leptospira Spp. through TaqMan Polymerase Chain Reaction Targeting the LipL32 Gene. *Diagn. Microbiol. Infect. Dis.* **2009**, *64*, 247–255. [CrossRef]
27. Kumar, S.; Stecher, G.; Suleski, M.; Sanderford, M.; Sharma, S.; Tamura, K. MEGA12: Molecular Evolutionary Genetic Analysis Version 12 for Adaptive and Green Computing. *Mol. Biol. Evol.* **2024**, *41*, msae263. [CrossRef]
28. Ruiz-Aravena, M.; McKee, C.; Gamble, A.; Lunn, T.; Morris, A.; Snedden, C.E.; Yinda, C.K.; Port, J.R.; Buchholz, D.W.; Yeo, Y.Y.; et al. Ecology, Evolution and Spillover of Coronaviruses from Bats. *Nat. Rev. Microbiol.* **2022**, *20*, 299–314. [CrossRef]
29. Hemnani, M.; Da Silva, P.G.; Thompson, G.; Poeta, P.; Rebelo, H.; Mesquita, J.R. Detection and Prevalence of Coronaviruses in European Bats: A Systematic Review. *EcoHealth* **2024**, *21*, 125–140. [CrossRef]
30. Petrović, T.; Lupulović, D.; Paunović, M.; Vidanović, D.; Lazić, G.; Samojlović, M.; Lazić, S. Detection and Typing of Coronaviruses in Bats in Serbia. In Proceedings of the 12th International Congress for Veterinary Virology, Ghent, Belgium, 20–23 September 2022.
31. Woo, P.C.Y.; Lau, S.K.P.; Lam, C.S.F.; Lau, C.C.Y.; Tsang, A.K.L.; Lau, J.H.N.; Bai, R.; Teng, J.L.L.; Tsang, C.C.C.; Wang, M.; et al. Discovery of Seven Novel Mammalian and Avian Coronaviruses in the Genus Deltacoronavirus Supports Bat Coronaviruses as the Gene Source of Alphacoronavirus and Betacoronavirus and Avian Coronaviruses as the Gene Source of Gammacoronavirus and Deltacoronavirus. *J. Virol.* **2012**, *86*, 3995–4008. [CrossRef] [PubMed]
32. Decaro, N.; Lorusso, A. Novel Human Coronavirus (SARS-CoV-2): A Lesson from Animal Coronaviruses. *Vet. Microbiol.* **2020**, *244*, 108693. [CrossRef]
33. Korneenko, E.V.; Samoilov, A.E.; Chudinov, I.K.; Butenko, I.O.; Sonets, I.V.; Artyushin, I.V.; Yusefovich, A.P.; Kruskop, S.V.; Sinitsyn, S.O.; Klyuchnikova, E.O.; et al. Alphacoronaviruses from Bats Captured in European Russia in 2015 and 2021 Are Closely Related to Those of Northern Europe. *Front. Ecol. Evol.* **2024**, *12*, 1324605. [CrossRef]
34. Ohlopkova, O.V.; Popov, I.V.; Popov, I.V.; Stolbunova, K.A.; Stepanyuk, M.A.; Moshkin, A.D.; Maslov, A.A.; Sobolev, I.A.; Malinovkin, A.V.; Tkacheva, E.V.; et al. Detection and Phylogenetic Analysis of Alphacoronaviruses in Bat Populations of Rostov and Novosibirsk Regions of Russia, 2021–2023. *Microbiol. Res.* **2024**, *16*, 3. [CrossRef]
35. Corduneanu, A.; Zając, Z.; Kulisz, J.; Wozniak, A.; Foucault-Simonin, A.; Moutailler, S.; Wu-Chuang, A.; Peter, Á.; Sándor, A.D.; Cabezas-Cruz, A. Detection of Bacterial and Protozoan Pathogens in Individual Bats and Their Ectoparasites Using High-Throughput Microfluidic Real-Time PCR. *Microbiol. Spectr.* **2023**, *11*, e01531-23. [CrossRef]
36. Nikolić, M. Pasteur's Institutes in Serbia. *Bull. Inst. Hyg.* **1954**, *III*, 53–72.
37. Stankov, S.; Lalošević, D.; Fooks, A.R. History of Rabies Incidence and Rabies Control in Serbia in Support of the Zero by 2030 Campaign to Eliminate Dog-Mediated Human Rabies. *Viruses* **2021**, *14*, 75. [CrossRef] [PubMed]
38. Goletic, S.; Goletic, T.; Omeragic, J.; Supic, J.; Kapo, N.; Nicevic, M.; Skapur, V.; Rukavina, D.; Maksimovic, Z.; Softic, A.; et al. Metagenomic Sequencing of Lloviu Virus from Dead Schreiber's Bats in Bosnia and Herzegovina. *Microorganisms* **2023**, *11*, 2892. [CrossRef]
39. Kemenesi, G.; Tóth, G.E.; Mayora-Neto, M.; Scott, S.; Temperton, N.; Wright, E.; Mühlberger, E.; Hume, A.J.; Suder, E.L.; Zana, B.; et al. Isolation of Infectious Lloviu Virus from Schreiber's Bats in Hungary. *Nat. Commun.* **2022**, *13*, 1706. [CrossRef]
40. Szentivanyi, T.; McKee, C.; Jones, G.; Foster, J.T. Trends in Bacterial Pathogens of Bats: Global Distribution and Knowledge Gaps. *Transbound. Emerg. Dis.* **2023**, *2023*, 9285855. [CrossRef]

41. Fritschi, J.; Marti, H.; Seth-Smith, H.M.B.; Aeby, S.; Greub, G.; Meli, M.L.; Hofmann-Lehmann, R.; Mühldorfer, K.; Stokar-Regenscheit, N.; Wiederkehr, D.; et al. Prevalence and Phylogeny of Chlamydiae and Hemotropic Mycoplasma Species in Captive and Free-Living Bats. *BMC Microbiol.* **2020**, *20*, 182. [CrossRef] [PubMed]
42. Mascarelli, P.E.; Keel, M.K.; Yabsley, M.; Last, L.A.; Breitschwerdt, E.B.; Maggi, R.G. Hemotropic Mycoplasmas in Little Brown Bats (*Myotis lucifugus*). *Parasites Vectors* **2014**, *7*, 117. [CrossRef] [PubMed]
43. Vengust, M.; Knapic, T.; Weese, J.S. The Fecal Bacterial Microbiota of Bats; Slovenia. *PLoS ONE* **2018**, *13*, e0196728. [CrossRef]
44. Matei, I.A.; Corduneanu, A.; Sándor, A.D.; Ionică, A.M.; Panait, L.; Kalmár, Z.; Ivan, T.; Papuc, I.; Bouari, C.; Fit, N.; et al. Rickettsia Spp. in Bats of Romania: High Prevalence of Rickettsia Monacensis in Two Insectivorous Bat Species. *Parasites Vectors* **2021**, *14*, 107. [CrossRef]
45. Mühldorfer, K. Bats and Bacterial Pathogens: A Review. *Zoonoses Public Health* **2013**, *60*, 93–103. [CrossRef]
46. Allocati, N.; Petrucci, A.G.; Di Giovanni, P.; Masulli, M.; Di Ilio, C.; De Laurenzi, V. Bat-Man Disease Transmission: Zoonotic Pathogens from Wildlife Reservoirs to Human Populations. *Cell Death Discov.* **2016**, *2*, 16048. [CrossRef]
47. Di Bella, S.; Giacchino, I.; Blanda, V.; Gucciardi, F.; Scibetta, S.; La Russa, F.; Lastra, A.; Purpari, G.; Grasso, R.; Spena, M.T.; et al. Zoonotic Bacteria and Vector-Borne Protozoa in Troglophilus Bat Colonies in Sicily (Southern Italy): A Biomolecular Survey. *Animals* **2025**, *15*, 488. [CrossRef]
48. Verde, R.S.; Di Azevedo, M.I.N.; Dias, D.; Tavares de Freitas, T.P.; Carvalho-Costa, F.A.; Bonvicino, C.; Lilenbaum, W.; D'Andrea, P.S.; Medeiros, L.S. Bat-Associated Pathogenic *Leptospira* spp. from Forest Fragments in Southwestern Brazilian Amazonia. *Transbound. Emerg. Dis.* **2024**, *2024*, 6633866. [CrossRef]

veterinary sciences

MDPI

Article

Comparative Mutational Analysis and the Glycosylation Patterns of a Peruvian Isolated Avian Influenza A Virus H5N1: Exploring Possible Viral Spillover Events Within One Health Approach

Sandra Landazabal-Castillo *, Lucero Alva-Alvarez, Dilan Suarez-Agüero, Enrique Mamani-Zapana and Egma Mayta-Huatuco

Molecular and Clinical Virology Laboratory, National University of San Marcos, Lima 15081, Peru; lucero.alva@unmsm.edu.pe (L.A.-A.); dilan.suarez@unmsm.edu.pe (D.S.-A.); emamaniz@unmsm.edu.pe (E.M.-Z.); emaytah@unmsm.edu.pe (E.M.-H.)

* Correspondence: sandra.landazabal@unmsm.edu.pe; Tel.: +51-983413000

Simple Summary: Influenza A virus is a pathogen of significant global concern for public health and wildlife conservation. A deeper understanding of its mutational/glycosylation profile is essential, given its broad host range capabilities. This study provides the comparative variations found in all segmented proteins of the H5N1 viruses analyzed, some of the findings must be highlighted within the One Health framework.

Abstract: (1) Background: The ongoing panzootic of highly pathogenic avian influenza virus (HPAIV) of subtype H5N1, clade 2.3.4.4b, has decimated wild/domestic birds and mammals' populations worldwide with reports of sporadic cases in humans. (2) Methods: This study aimed to compare the mutational profile of H5N1 avian Influenza virus isolated from a Peruvian natural reserve, with recent data from other related international studies made in human and different species of domestic and wild birds and mammals. Briefly, the near complete protein sequences of the Influenza virus coming from a *Calidris alba* were analyzed at a multisegmented level, together with 55 samples collected between 2022 and 2024 in different countries. Moreover, the glycosylation patterns were also predicted in silico. (3) Results: A total of 603 amino acid changes were found among H5N1 viruses analyzed, underscoring the detection of critical mutations HA:11I, HA:211I, HA:336T, HA:492D, HA:527I, NA:10T, NA:269L, NA:405T, NP:377N, PA:57R, PA:68S, PA:322V/L, PA:432I, PB2:539V, PB1:207R, PB1:375N, PB1:264D, PB1:429R, PA-X:250Q, PB1-F2:65R, and PB1-F2:42Y, as well as PA:13V, PA-X:13V, PA20T, PA-X:20T, PA:36T PA-X:36T, PA:45S, PA-X:45S, PA:57Q, PA-X:57Q, PA:61I, PA-X:61I, PA:68S, PA-X:68S, PA:70V, PA-X:70V, PA:75Q, PA-X:75Q, PA:85T, PA-X:85T, PA:86I, PA-X:86I, PA:100I, PA-X:100I, PA:142E, PA-X:142E, PA:160E, PA-X:160E, PA:211I, PA-X:211Y, among others, considered of importance under the One Health perspective. Similarly, changes in the N-linked glycosylation sites (NLGs) predicted in both HA and NA proteins were found, highlighting the loss/acquisition or changes in some NLGs, such as 209NNTN, 100 NPTT, 302NSSM (HA) and 70NNTN, 68NISS, and 50NGSV (NA). (4) Conclusions: This study provides our understanding about the evolution of current Influenza A viruses H5N1 HPAIV circulating globally. These findings outline the importance of surveillance updating mutational profiles and glycosylation patterns of these highly evolved viruses.

Keywords: One Health; Influenza A viruses; H5N1; mutational analysis; N-linked glycosylations (NLGs) profile

https://doi.org/10.3390/vetsci12040392

1. Introduction

Influenza A virus (IAV) is an enveloped, negative-sense, single stranded, and eight segmented RNA virus, belonging to the genus *Alphainfluenza*, family *Orthomyxoviridae*, which is a major threat within a One Health framework of global concern for public health and wildlife conservation due to its highly contagious spread and the devastating impact it has caused in breeding colonies of different species of domestic and wild birds and mammals globally [1,2]. The ultimate record of these viruses across the Americas, spreading from Canada to the Antarctic and sub-Antarctic regions of South Georgia, the Sandwich Islands, and the Falkland Islands, has decimated animal populations, including places where high pathogenicity avian Influenza virus (HPAIV) had never previously been detected [3,4]; in regions of Peru, Chile, Argentina, Uruguay, and Brazil, the spreading of the virus has particularly affected marine mammal populations with increasing evidence of mammal–mammal transmission, and comprises mutations that were not reported in H5N1 viruses in other regions [5–7]. Likewise, the emerging infections of H5N1 avian origin virus reported in cattle [8,9], wild carnivores [10,11], domestic mammals [12,13], as well as reports of human infections [14] highlighted the characteristics of these exceptionally evolved viruses and the risk of possible spillover events.

Furthermore, the continuous outbreaks of Influenza H5N1 HPAIV clade 2.3.3.4b viruses never seen before worldwide represents a serious threat, given the ideal conditions to share key mutations between different species, including hosts with the potential to act as mixing vessels because they carry both avian and human-type receptors. The genetic variations acquired by antigenic drift (minor mutational changes over time) or shift (abrupt major change in the virus where genome segments may be rearranged to another subtypes), have increased the opportunities of viral evolution, and the emergence/reemergence of new virus strains. Nowadays, the risk of H5N1 host tropism broadening with the possibility of a panzootic/pandemic is looming; therefore, a primordial concern centers around the virus's capabilities for genetic changes or reassortment events [15]. Hence, active surveillance of influenza mutational profiles is required, providing the opportunity to assess the risk posed by the different subtypes/clades/genotypes, that may enhance the mammalian adaptation, spillover events, and the likelihood of reassortment events, particularly H5N1 clade 2.3.4.4.b virus, quantifying their zoonotic potential.

A risk assessment framework was recently made to characterize the variations related to phenotypic traits for the adaptation to new hosts based in epidemiological and spatio-temporal data analysis of several Influenza viruses. A total of 592 mutations where identified, where 34 of them were associated with an increase in zoonotic potential, and five phenotypic traits (haemagglutinin stability, neuraminidase specificity, major receptor specificity, improved polymerase activity, and immune evasion). It was featured that H5 viruses of clade 2.3.4.4.b have acquired the highest number of zoonotic traits (up to four), in the European Union, during the last three years; emphasizing that there is still limited evidence, poor information available or not investigated for further mutations [16]. In addition, though the information related to certain mutations, and these biological properties, is scarce, it has been observed that some adaptational variations to new hosts can be provided through mutations [17,18]; such as NA:L204M, which enhanced viral replication, receptor binding, and virulence; PB2-E627K, which increased viral replication in mammals or HA:156A/V, which increased affinity for α2–6 the human type receptor. These are only examples of the high plasticity of these segmented viruses. Likewise, it is well known that the lofty variability of Influenza viruses presented over time during outbreaks, such in the case of two Influenza A/H1N1 pandemic viruses from 1918, that compared to the reference genome A/Brevig Mission/1/1918 (H1N1), a total of 76 SNPs from formalin-fixed, paraffin-embedded lung tissues were identified, of which 46 encode non-synonymous (NS) changes [19].

Additionally, key mutations that occurred in HA and NA proteins can vary the biophysical properties of Influenza viruses when N-linked glycosylation motifs are introduced/removed, which can facilitate immune evasion or increase the receptor binding affinity to maintain viral fitness [20]. It is well known that variable degrees of glycosylation confer capabilities to the globular head and stem of the HA ectodomain to bind host-derived glycans and hide or expose the functional region, a process required to initiate the cell viral entry [21]; moreover, glycosylation of NA protein has been related to changes in protein folding, stability, solubility, and viral budding, as well as being related to the neurovirulence of Influenza viruses [22,23]. In addition, the glycosylation dynamics in the current H5N1 Influenza viruses are not yet fully understood at all.

Given the critical role of mutations and glycosylation patterns in the dynamics of Influenza virus and its adaptive evolution, this study describes the comparative results obtained between several H5N1 Influenza virus genomes, mammal and non-mammal, within One Health approach.

2. Materials and Methods

2.1. Sample Collection

One duplicate sample oropharyngeal swab was collected from a *Calidris alba* with slowness fly and stagnation from Pantanos of Villa a Peruvian National Reserve during the period of time from March–April 2023. The material was transported and preserved in a triple packaging system to perform the molecular detection of Influenza A viruses following the standard diagnostic procedures by amplification [24].

2.2. Molecular Detection

Viral RNA was extracted using the Viral Nucleic acid extraction kit II (Geneaid, New Taipei, Taiwan) according to the manufacturer's instructions in a class 2A biological safety cabinet (Biobase, Jinan, China). The extracted RNA was subjected to purity measurement using a spectrophotometer DS11 instrument (Denovix, Wilmington, USA) and Influenza virus detection was performed using a high-resolution melting analysis (HRM) procedure to target the M and HA gene following the standard procedures of World Organization of Health [25].

2.3. Whole-Genome Sequencing

Whole-genome sequencing (WGS) of amplicons per each genome segment of the isolate H5N1 virus were obtained using Mytaq® Red DNA Polymerase kit (Meridian Bioscience, Cincinnati, OH, USA) and the sequencing was performed using Illumina next generation sequencing (NGS) technology (Miseq system with a 250-cycle paired-end). The reads were analyzed according to the tools described in our previous publication [26].

2.4. Data Sets

Full-length or near full-length protein sequences of 55 highly pathogenic avian Influenza A viruses (HPAIV) H5N1 were downloaded from the National Center for Biotechnology Information (NCBI), together with data from the Peruvian isolate H5N1.A/Calidris alba/Lima/Villa01/2023, within One Health perspective. The sample list included available data from outbreaks in 2022 and 2024, in the Americas region mainly and considered of importance to be included in the analysis. The criteria for data selection were geographic; host diversity (representative species of mammals/birds, some without a previous report of infection with H5N1 virus of avian origin); availability of complete/near complete protein sequences; data quality reports, and related recent publications. All sequences obtained from GenBank are listed in Supplementary S1, Table S1. All Influenza virus proteins were included in the perusal: PB2, PB1, PA, HA, NP, NA, M, and NEP (NS2), and the accessible information of Influenza virus accessory proteins were taken into special consideration.

2.5. Mutational Analysis and Genotype Identification

The data obtained at a multispecies level were analyzed in the search of amino acid changes, segment by segment manually and in silico, using Clustal Omega v12.4 [27], along with the command-line tool FluMut [28] and the web application Flusurver [29]. All available sequences of PB2, PB1, PA, HA, M, NP, NA, and NEP (NS2) were compared, including the accessory proteins PA-X, PB2-F1, and NS1. Genotype was determined using GenoFLU (https://github.com/USDA-VS/GenoFLU (accessed on 26 January 2025)) [30].

2.6. Prediction of Potential N-Glycosylation Sites in HA and NA Influenza A H5N1 Viruses

Potential N-linked glycosylation sites (NLG) of the sequences HA and NA proteins of the analyzed H5N1 viruses were predicted and compared with the reference genome (A/goose/Guandong/1996(H5N1) using the high prediction accuracy NetN-Glyc 1.0 server [31], detecting the Asn-X-Ser/Thr sequons (where X is any amino acid except proline) [32], and measures of prediction confidence >0.5 were taken into consideration as a threshold.

3. Results

3.1. Mutational Manual Analysis

Through mutational manual analysis of the selected 55 samples of H5N1 Influenza viruses, coming from both birds and mammals, a total of 603 amino acid changes were identified, distributed throughout the 8 protein segments PB1, PB2, PA, HA, M, NA, NP, NEP, and the 4 accessory proteins M2, NS1, PB1-F2, and PA-X; where 76 mutations have been detected in the HA protein; 72 mutations in the NA protein; 22 mutations in M1; 11 mutations in M2; 22 mutations in NP protein; 94 mutations in NS1 protein; 39 mutations in NEP (NS2) protein; 56 mutations of PB2 protein; 64 mutations in PB1 protein; 48 mutations PB1-F2; 30 mutations in PA-X; and 75 mutations in PA protein (Figure 1). The complete mutational details can be observed in Supplementary S1 and S2.

It is noteworthy that the proportion of individual/group mutations detected per each protein segment, varied notoriously as follows: individual mutations (those found only in 1–3 of the samples evaluated) were observed more in PA, PB1, PB2, NP, NA, and PA-X proteins; while group mutations (those identified in more than 4 samples), were detected in larger quantities in PB1-F2 and PB1 proteins. In addition, more mutations in "all" samples analyzed, were found in M1 and NS1 proteins (S1 (Table S2)).

Furthermore, considering the size of each virus protein segment, the detection of more quantity of amino acid variations in the accessory proteins PB1-F2, NS1, and NEP, particularly in the "PB1-F2 protein" was conspicuous. Viruses such as A/LOU/WT, polar bear/ALK, and harbor-seal/ME, along with certain birds species shared 23 variations in PB1-F2 protein; likewise, they also had in common 7 mutations in PB1, 4 mutations in HA, 2 mutations in NS1, 1 mutation in PA-X, as well as group mutations PB1:375N, HA:492D, PB1-F2:22E, PB1-F2:90N, NS1:83P, and PA-X:215N. Similarly, a considerable number of individual mutations were found in proteins, NA (19 variations detected in A/LOU/WT, 3 in polar-bear/ALK), and PA (13 variations detected in A/LOU/WT, 6 in polar-bear/ALK). Additionally, the individual mutations NA:44N, NA:45H, NA:48T, NA:53V, NA:62I, NA:81D, NA:82P, NA:84A, NA:234I, NA:286S, NA:288V, NA:399L, and PA:201I/T, PA:211I, PA:322L, PA:399V, PA:626R were also found in these isolates. Keeping in mind that these specimens are the most recent isolates and considering the host species, it should be considered under the One Health perspective (Tables 1 and 2).

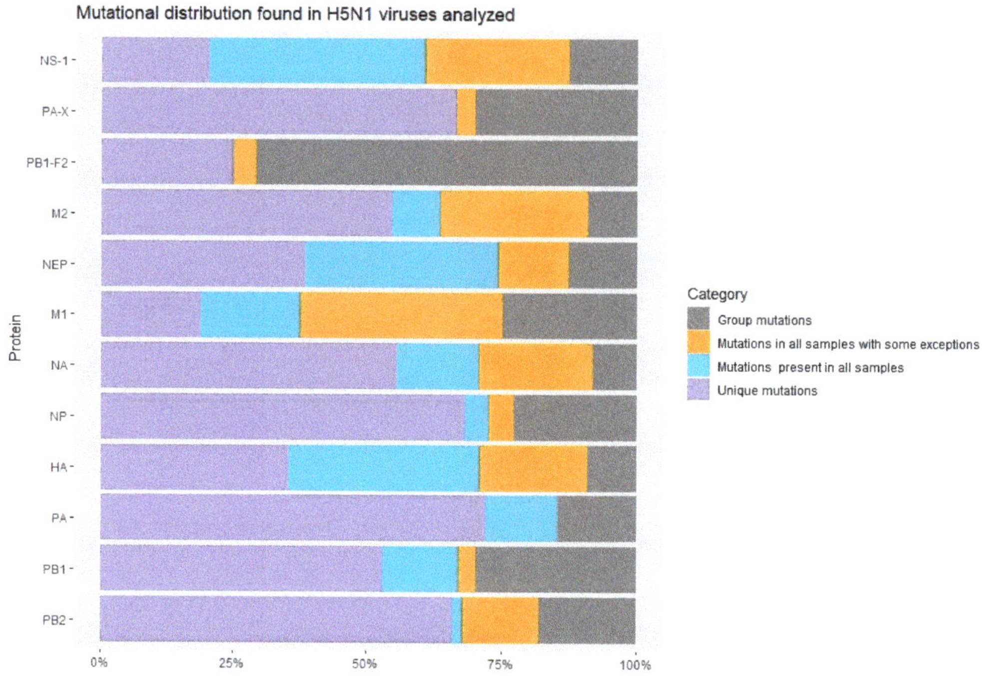

Figure 1. Mutational distribution found in the H5N1 viruses analyzed. The mutations were classified into 4 categories: unique mutations (mutations found only 1–3 samples), group mutations (mutations found in more of 4 samples), mutations present in all samples, and mutations found in almost all samples with some exceptions.

Table 1. Unique mutations (only detected in 1–3 samples) are considered of interest in H5N1 viruses identified per each protein segment.

Mutation	Mutation to Highlighting
HA:9V, HA:87T, HA:99S/D, HA:225M, HA:248L, HA:259C, HA:277Y, HA:285E, HA:316E, HA:324T, HA:473K, HA:493K, HA:531L	HA:10T, HA:170D, HA:104G, HA:147M, HA:152S, HA:226T, HA:304N, HA:310V, HA:336N/R, HA:520R/N
NA:20A/I, NA:216V, NA:217R, NA:223T, NA:284N, NA:308R/K, NA:340Y/F, NA:364N, NA:442I	NA:23V, NA:44N, NA:45H, NA:48T, NA:53V, NA:62I NA:67I, NA:74L, NA:75I, NA:81D; NA:82P, NA:84A, NA:90P, NA:155H, NA:221S, NA:234I, NA:237F, NA:241I, NA:254R, NA:257R, NA:286S, NA:288V, NA:329S, NA:374V, NA:399L, NA:432R, NA:436V
M1:55M, M1:125T, M1:191H, M1:218A, M1:236K	
M2:12R, M2:21G, M2:28T, M2:52S	M2:27A

Table 1. *Cont.*

Mutation	Mutation to Highlighting
PA:42V, PA:59K/G, PA:118U, PA:184S, PA:190F, PA:207V, PA:213K, PA:272N, PA:269K, PA:323I, PA:330V, PA:382G, PA:423T, PA:425F, PA:523L, PA:561V, PA:581I, PA:621V, PA:664R, PA:688G	PA:13V, PA:36T, PA:45S, PA:68S, PA:75Q, PA:86I, PA:100I, PA:142E, PA:201I/T, PA:211I, PA:322L/V, PA:336M, PA:348L, PA:351G, PA:354F, PA:388G, PA:399V, PA:404S, PA:459V, PA:465M/T, PA:486M/L, PA:489S, PA:538G, PA:545V, PA:614D/S, PA:626R, PA:655F
PA-X:42V, PA-X:52D, PA-X:62T, PA-X:118V, PA-X:122I, PA-X:184N, PA-X:190F, PA-X:207L,	PA-X:20T, PA-X:36D/T, PA-X:68S, PA-X:70V, PA-X:75Q, PA-X:86I, PA-X:142E, PA-X:160E, PA-X:195K, PA-X:211Y, PA-X:250P,
PB2:191G, PB2:199T, PB2:292V, PB2:339R, PB2:444G, PB2:451V, PB2:452V, PB2:453S, PB2:472D, PB2:560M, PB2:575V, PB2:639S, PB2:660R, PB2:679S, PB2:683A, PB2:684S, PB2:697M, PB2:711S	PB2:9N, PB2:79G, PB2:152V, PB2:190R, PB2:251K, PB2:255A, PB2:274V, PB2:346A, PB2:353R, PB2:532L, PB2:539V, PB2:596A, PB2:663R, PB2:666I, PB2:667I, PB2:670R, PB2:677K, PB2:680G, PB2:715S
PB1:11R, PB1:14V, PB1:51E, PB1:53E, PB1:121N, PB1:147V, PB1:176T, PB1:321I, PB1:339V, PB1:348V, PB1:371D, PB1:383G, PB1:455D, PB1:394S, PB1:431H, PB1:512L, PB1:576M, PB1:584H, PB1:657H, PB1:719I, PB1:739D	PB1:40I, PB1:171A, PB1:211K, PB1:291A, PB1:372I, PB1:384P/T/A, PB1:390G, PB1:533S, PB1:621K, PB1:660I, PB1:738G
NS1:66D, NS1:81V, NS1:88H, NS1:129T, NSI:210R, NSI:202T, NS1:217T, NSI:213L, NSI:219E	NSI:36I, NS1:67G/Q, NS1:75G, NS1:76A, NS1:77R, NS1:136M, NS1:193Q, NSI:201Y
NEP:27G, NEP:52V, NEP:56Y, NEP:61K, NEP:64T, NSI:76M, NEP:77K, NEP:85Q NEP:81G, NEP:82E	NEP:36V, NEP:60N, NEP:63E, NEP:89T/V
PB1-F2:11R/L, PB1-F2:29R, PB1-F2:35L, PB1-F2:41L, PB1-F2:69L, PB1-F2:78R, PB1-F2:79Q, PB1-F2:90I	PB1-F2:39T, PB1-F2:57Y, PB1-F2:73E
NP:41V, NP:190A, NP:221K, NP:234S, NP:253V, NP:323S, NP:363I	NP:48R, NP:63T, NP:119T/V, NP:230L, NP:318L, NP:411A, NP:425V

The complete list of mutations per each segment with details of the host can be seen in Supplementary S1. Mutations highlighted in blue were observed in both bird and mammal species or by the host which was found.

Moreover, unique and/or group mutations stands out among other North/South America specimens, highlighting that many of them have also been found in both mammals and birds, such in case of individual variations shared found in North American isolates: A/CA, emu/CA (HA:104G, HA:336N, PA:68S, PA:486M, PA:655F, PA-X:68S, PB2:670R, NS1:67G, NP:119V); cattle/TX (M2:27A, PA:36T, PA:404S, PA-X:36T, PB1:384T, NEP:60S); South American Chilean-dolphin, elephant-seal/ARG, and terns/ARG (NS1:26K, NSI:226T, NP:119T, PA:57Q, PA-X:57Q, PA:86I, PA-X:86I, PA:336M, PA-X:20T, PB2:152V, PB1:40I, PB1:548F, PB1:515A, PB1:621K. Similarly, other specimens also had their own particular mutations, such as in South America: Panthera-leo/PER (HA:310V, PA:45S, PB2:190R, NEP:89V, NP:425V), Calidris-alba/LIM (HA:201R, NA:442I, M2:52S, PB2:679S, NS1:213L, NEP:56Y, NP:190A); black-necked-swan/UGY (PA:425F, PB2:199T, NP:323S, PB1-F2:90I); Andean-guayata/ARG (HA:9V); and in North America: peregrine-falcon/NY (NA:223T, NA:237F, M2:19Y, PA-X:118V, PA-X:207L, PA-X:250P, PB2:472D), vulture/FL (HA:87T, HA:102T, PA:59G, PA:272N, PA-X:59K, PB2:453S, PB1:431H, NEP:82E, PB1-

F2:41L, PB1-F2:69L); harbor-seal/ME (HA10T, HA:152S, HA:226T, PA:465T, PA-X:70V, PB2:79G, PB2:715S, PB1:176T, PB1:372I, PB1:660I, NS1:67Q, PB1-F2:73E), goat/MI (HA:520R, PA:614S, PB2:274V, PB2:346A, PB2:353R, PB2:663R, PB2:667I, PB1:211K, NS1:36I, NS1:136M, NS1:201Y); raccoon/IA (M2:28T, PA:100I, PB2:539V); house-mouse/NM (NA:254R, PA:13V, PB1:384P, NS1:77R); alpaca (PA:142E, PA-X:142E), and red-fox/MI (PB2:596A). Additionally, the group mutations PB1:515A, NS1:26K, PA-X:86I and PA-X:57Q were detected in the South American mammals and birds analyzed.

Table 2. Mutations identified in H5N1 viruses per samples group (4 or more isolates).

Mutation	Mutation to Highlighting
HA:242I, HA:504Y	HA:11I, HA:52A, HA:211I, HA:492D, HA:527I
NA:71S, NA:321I	NA:6R, NA:10T, NA:70N, NA:405T
M1:82S, M1:227T	
M2:88N/D	
NP:52H, NP:105M, NP:293K, NP:482N	NP:377N
PB2:154F, PB2:362G, PB2:441N, PB2:495I, PB2:631L, PB2:649I, PB2:676A	PB2:58A, PB2:109I, PB2:139I,
PB1:179I, PB1:587P, PB1:646I	PB1:16D, PB1:154S, PB1:172D, PB1:207R, PB1:215K, PB1:264D, PB1:375N, PB1:378M, PB1:399D, PB1:429R, PB1:430K/E, PB1:515A, PB1:548F, PB1:614D, PB1:694S
PB1-F2:30L, PB1-F2:31E, PB1-F2:44R, PB1-F2:46T, PB1-F2:47S, PB1-F2:48R, PB1-F2:50G, PB1-F2:54K, PB1-F2:55I, PB1-F2:66S, PB1-F2:68I, PB1-F2:70G	PB1-F2:4G, PB1-F2:7I/T/M, PB1-F2:8Q, PB1-F2:12S PB1-F2:17S, PB1-F2:18T, PB1-F2:20R, PB1-F2:21R, PB1-F2:22E, PB1:36T, PB1-F2:40G, PB1-F2:42Y, PB1-F2:49A, PB1-F2:56A, PB1-F2:57C/F, PB1-F2:58W, PB1-F2:65R, PB1-F2:75L, PB1-F2:82S, PB1-F2:84S, PB1-F2:90N
NS1:7S/L, NS1:21Q, NS1:88C, NS1:189N	NSI:21R, NS1:26K, NSI:53G, NS1:83P, NS1:87S, NS1:116C/S/N, NS1:147I, NS1:226T
NEP:7V/S, NEP:31V, NEP:67G/E	
PA:113R, PA:237K/A, PA:272E, PA:277P, PA:479E, PA:558L	PA:57R, PA:219I, PA:432I, PA:497R,
PA-X:113R, PA-X:215L,	PA-X:57Q, PA-X:61I, PA-X:85T, PA-X:193S, PA-X:245N

Mutations highlighted in blue were observed in both bird and mammal species or by the host which was found.

Further, certain samples accumulated a greater quantity of individual/group amino acid changes in different protein segments such as in case of birds coming from Brazil (Numida/BR, Procellaria/BR mainly) and the specimens coming from Bangladesh, Japan, Egypt, and Germany included within the present study. (Tables 2 and 3).

In addition, a number of mutations were found in "all" specimens, (concerning the reference genome) (Table 3), and in other cases some specimens had mutations shared in almost all samples with little exceptions detected (Table 4). Be aware that several of these mutational changes could have been transmitted over time at a multispecies level, bringing them with/without a specific trait, effect, or significance; nonetheless, these were common findings even in specimens far away from our geographical region.

Table 3. Mutations detected in all H5N1 viruses analyzed.

Protein	Mutation
HA	110S, 139P, 142E, 143T, 154Q, 157P, 171D, 208K 234Q, 239R
NA	46P, 76A, 78Q, 99I, 100Y, 258I, 289M, 366S, 382E, 418M, 434N
M1	140A, 144L, 165I,
M2	18N
NP	136L
PB2	699K, 741S
PB1	177E, 478S, 490F, 535I, 536N, 558T, 598P, 609Y, 610C
PB1-F2	No PB1-F2 sequence in the reference genome
NS1	6I, 18V, 22F, 23S, 24D, 25Q, 27L, 28C, 54I, 60A, 73S, 84V, 94T, 95L, 112A, 114G, 117I, 127R, 137L, 140Q, 146L, 153E, 158G, 161S, 163L, 170T, 180V, 191T, 194V, 197T, 198L, 205S, 206S, 211R, 221K, 224R, 225T
NEP (NS2)	6V, 14M, 22G, 26E, 37S, 40L, 48A, 49V, 68Q, 83V, 86R, 88K, 100M, 111Q
PA	63V, 129I, 212C, 228N, 361K, 536K, 544E, 585L, 586L, 716R
PA-X	no PA-X sequence in the reference genome
PB2	355R, 699K

The comparisons were performed with the reference genome A/goose/Guandong/1996(H5N1), which no have sequences available to the accessory proteins PB1-F2 and PA-X. Mutations highlighted in blue were observed in both bird and mammal species or by the host which was found.

Table 4. Mutations found in almost all H5N1 viruses with some exceptions.

Protein	Mutation
PB1	59S, 75D
PB2	334S, 340R, 463V, 464M, 471T, 478I, 590G, 616V
NP	450N
M2	28I, 51V, 61G
M1	85S, 87T, 101R, 200V, 230R, 232D
HA	120M, 131L, 199N, 201E, 205N, 226A/T, 336S/N, 527V/I, 549M
NA	8T, 20V/A, 44Y/N, 81T/D, 155Y, 188I, 269M, 287D, 340S, 336S, 338M, 339P, 340S, 395E, 460G
NEP	63A, 64K, 81E, 85H, 89I
PA-x	252R/K
PB1-F2	11Q, 12L/S
NS1	44R, 55E, 56T, 59R, 63Q, 70E/G, 71E, 74D, 90L, 111V/A, 118R, 139D, 145I, 166L, 171D, 192V, 204R, 207N, 209D, 213P, 226T

Mutations highlighted in blue were observed in both bird and mammal species or by the host which was found.

In general, during the mutational scanning, certain mutations were acquired in all samples tested, whilst others were only detected in certain groups of them. But, evidently, there was an existing "different" mutations distribution between the North American/South American isolates. Also, among the North American samples, two large groups with similar amino acid changes were distinguishable: a first large group with certain homogeneity in the mutations found, and the second smaller that included the most recent specimens that shared new mutations, of which there are no still studies on the possible biological function, such as PB1-F2:4G, PB1-F2:7I/T/M, PB1-F2:8Q, and PB1:12S among others (complete mutational scanning can be seen in Supplementary S2.

3.2. *FluMut Mutational Analysis*

Our mutational analysis, were compared with the results obtained with FluMut program, and a match in the detection of the next mutations/molecular markers was found as follows: HA:154N, HA:156A, M2:27I; NA:155H, NA:223T, NA:364N; NS1:53D, NS1:55E, NS1:66E, NS1:74N, NS1:205S, NS1:210R; NS2:48A; NP:41V, NP:105V; PA:63I, PA:142E, PA:190S, PA:497R; PB1:207R, PB1:375S, PB1:598L; PB1-F2:56A, PB1-F2:66S; PB2:9N, PB2:292V, PB2:339K, PB2:495I, PB2:590S, PB2:631L, PB2:676T, PB2:699R, PB2:715N; emphasizing that some of the mutations presented amino acid changes in the mutation designated, such as in the case of HA:154N (HA:154Q), or NA:53D (NA:53G), and taking into account, as previously mentioned, that certain mutations were found in all samples whilst others were only detected in individual samples. (Complete FluMut mutational analysis can be seen in Supplementary S3.

3.3. *Genotype Identification*

Four genotypes were found in the H5N1 viruses analyzed: B3.13(30.5%), B3.2(37.3%), B1.3(3.4%), A3(3.4%), A2(1.7%), B1.1(1.7%); and in (22%) of specimens the genotype was not assigned because not all the segments matched the total of segments in input file. (Complete Genoflu genotype determination can be seen in Supplementary S4.

3.4. *Glycosylation Patterns*

3.4.1. N-Linked Glycosylations (NLG) in the HA of Influenza H5N1 Viruses

A total of 7 N-linked glycosylations were found in HA protein of the reference genome (A/goose/Guandong/1996(H5N1) at positions 27 NSTE, 39 NVTV, 181 NNTN, 209 NPTT, 302 NSSM, 500 NGTY, and 559 NGSL, respectively, taking into consideration that 209 NPTT position had included a warning (Pro-X1), due the presence of a proline in the sequon. The complete list of glycosylation predictions is Supplementary S5. Particularly, the specimens that had an absent 209 NPTT NLG, had a change in the amino acid position 211 from T in the HA protein.

For instance, the isolates coming from North America had two characteristics (presence or absence of 209 NPTT NLG): first, 4 mammal specimens (2 A/LOU/WA, 1 harbor-seal/ME/ME, 1 polar bear/AK), and 6 birds ones (peregrine-falcon/NY, goose/AK, Northern-pintail/USA, chicken/JPN, duck/BD, white-tailed-eagle/JPN, turkey/GER and pintail/EG) had the predicted 209 NPTT glycosylation site, and showed 7 NLGs; second, 16 mammal samples (4 A/CA/MT/CO, 1 alpaca/ID, 3 bovine/TX/ID, 2 feline/TX/ID, 1 red-fox/MI, 2 goat/MI, 1 raccoon/USA, 1 house-mouse/NM, 1 domestic cat/TX), and 3 bird samples (Emu, chicken/Idaho and grackle/Texas) only had 6 NLGs predicted and had lost 209 NPTT. A comparison between the glycosylation sites predicted (HA protein) to the reference genome and the specimen cattle/TX can be observed in Figure 2.

In the case of mammal isolates coming from South America (elephant-seal/ARG, Panthera-leo/PE and dolphin/CHIL), as well as the 19 bird isolates (Andean-guayata/ARG, Sterna-hirundo/BR, Humboldt-penguin/CHIL, Procellaria-aequinoctialis/BR, Numida-meleagris/BR, cormorant/CHIL, belcher-gull/PE, wild-duck/CO, 2 pelecanus/PE, backyard-duck/UY, Fregata-magnificens/BR, chimango-caracara/CHIL, black-necked-swam/UY, royal-tern/AR, Calidris-alba/PE, gallus-gallus/PE, Thalasseus-acuflavidus/BR and South America-tern/ARG) all of these had 7 NLG, including 209 NPTT.

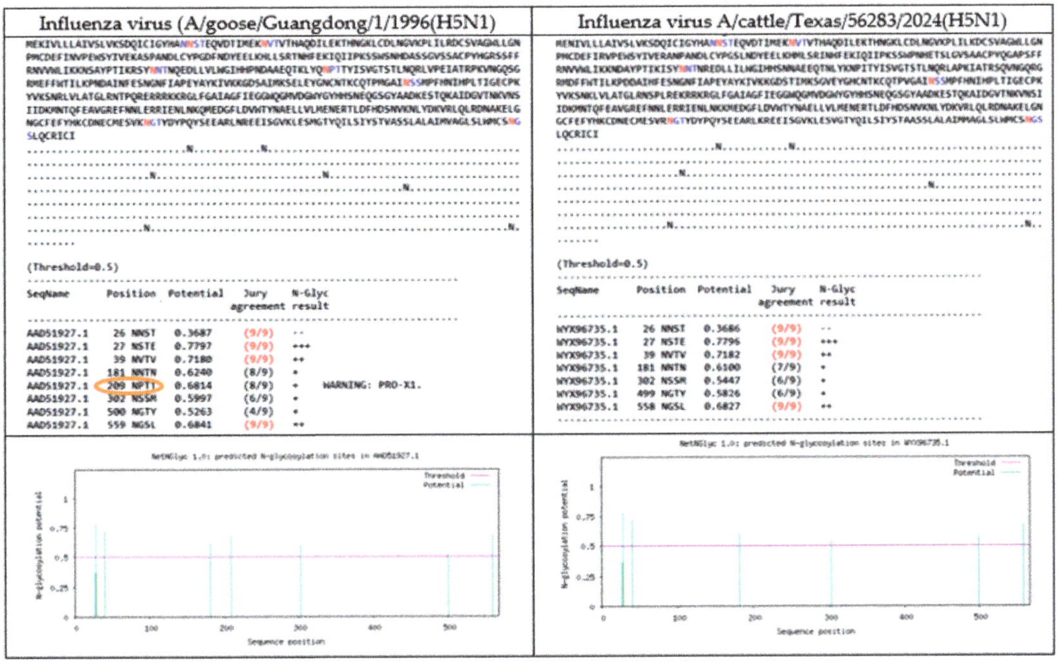

(A) **(B)**

Figure 2. Comparison of glycosylation sites for HA protein of H5N1 viruses: (**A**). Reference genome Influenza A virus (A/goose/Guangdong/1/1996(H5N1) had 7 NLGs predicted; (**B**). Influenza A virus/cattle/Texas/56283/2024, had 6 predicted NLGs, with substitution of 2 NLGs (499 NGTV and 558 NGSL) and is highlighthed the loss of the 209 NPTT site (bold in orange).

Likewise, the glycosylation predictor showed a substitution in the NLG site to 499 NGTY and 558 NGSL (from 500 NGTY and 559 NGSL of the reference genome) in all specimens analyzed. This ultimate characteristic coincided with the presence of a mutation in the amino acid sequence prior to the NGTY sequon, from K to R in position 499. In addition, a slight increase in the glycosylation potential was seen as 0.5263 (reference genome) versus 0.58 found in all samples analyzed.

Furthermore, 2 North American avian isolates (chicken/JPN and vulture/FL) had 8 predicted NLGs, with an extra NLG site as 100 NPTN, which also concurred with a mutation in the HA protein, 102 site from A to T amino acid; the feature was only shown in these two isolates. It should be noted that a 100 NPTN glycosylation prediction was found between samples very far away geographically.

In addition, two South American avian samples (Sterna-hirundo/BR and Humboldt-penguin/CHL) showed 6 NLGs with a loss of glycosylation site 302 NSSM. In both cases, a substitution was observed in the amino acid 304, where S was replaced by N. A comparison between the North American specimen (A/Vulture/Florida/2022(H5N1) with 8 NLGs glycosylation sites predicted (HA protein) versus the South American (A/Humboldt-penguin/CHI/2023(H5N1) with 6 NLGs is shown in Figure 3.

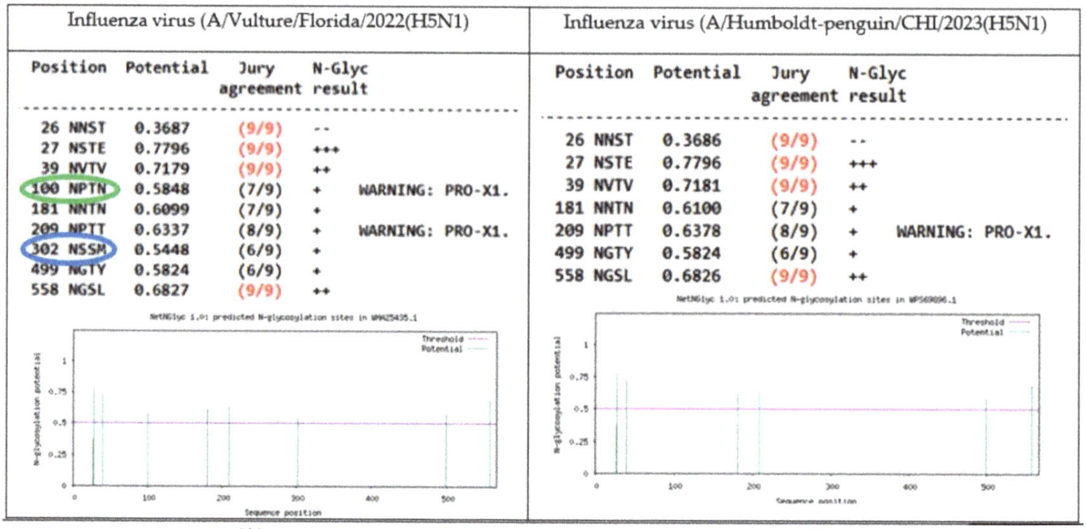

(A) (B)

Figure 3. Comparison of glycosylation sites for HA protein of H5N1 viruses: (**A**). Left. North American Influenza A virus (A/Vulture/Florida/2022(H5N1) had 8 predicted NLGs, with a gain of 100 NPTN site (green bold); (**B**). Right. South American Influenza A virus/Humboldt-penguin/CHI/56283/2023(H5N1) had 6 NGLs predicted with loss of 302 NSSM site (blue bold).

3.4.2. N-Linked Glycosylations in the NA of Influenza H5N1 Viruses

The predicted NLGs in NA protein of the reference genome (A/goose/Guandong/1996(H5N1) found 7 sites of glycosylation at 50 NQSI, 58 NNTW, 63 NQYT, 68 NISN, 88 NSSL, 146 NGTV, and 253 NGSC positions, and differences in the predicted glycosylation positions were found in both mammal and bird specimens.

First, the 3 South American mammals' isolates (elephant-seal/ARG, Panthera-leo/PE and dolphin/CHIL), as well as 17 South America avian isolates (Andean-guayata/ARG, Sterna-hirundo/BR, Humboldt-penguin/CHIL (Figure 3), Procellaria-aequinoctialis/BR, Numida-meleagris/BR, cormorant/CHIL, belcher-gull/PE, pelecanus/PE, backyard-duck/UY, Fregata-magnificens/BR, chimango-caracara/CHIL, black-necked-swam/UY, royal-tern/AR, Calidris-alba/PE, gallus-gallus/PE, thalasseus-acuflavidus/BR and South America-tern/ARG) presented the same 7 NLGs positions described for the reference genome. Nevertheless, the wild-duck/CO specimen had a substitution from 68 NISN to 70 NNTN. This same feature was also seen in the 2 avian isolates (duck/BD, white-tailed-eagle/JPN), and (polar-bear/ALK) a sample of mammal origin.

In the case of North American isolates, 2 avian samples (vulture/FL and Goose/AK, shared the same patron of glycosylation with NA reference genome, while 3 specimens (Emu/CA, grackle/TX, chicken/ID) had the amino acid change from N to S in position 68 (68NISN to 68NISS) of NA protein. It is worth highlighting that this characteristic was also observed in 11 North American mammal isolates (2 A/CA, and 1 A/CO, 1 domestic-cat/TX, 1 alpaca/ID, 3 bovine/TX/ID, 2 feline/TX and 1 racoon/USA), along with a substitution in the amino acid of the 67 position from V to I, just prior to the sequon location. Additionally, the potential of glycosylation predicted in this particular site was quite variable: from 0.73 in the reference genome, 0.67–0.68 in samples 68 NISN, such as (A/LOU/WT/MO, red-fox/MI and cat/TX isolates) or 0.71 in samples that presented the variation 68 NISS.

Furthermore, two recent North American mammal isolates (2 A/LOU/WT) had the unique characteristic 50 NQSV predicted, with a slight increase in glycosylation potential from 0.50 to 0.60 compared with other samples (50 NQSI), for instance the reference genome had a potential of 0.55 in this NLG predicted site. A comparative scheme of the predicted NLGs glycosylation sites (NA protein) of 4 representative isolates can be seen in Figure 4.

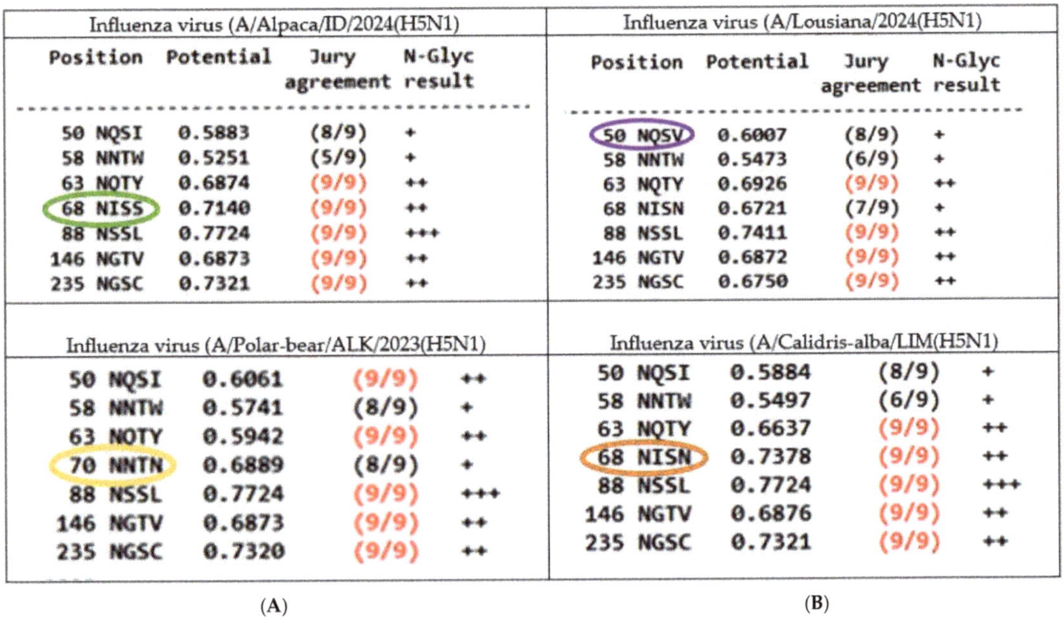

Figure 4. Comparison of glycosylation sites predicted for NA protein of H5N1 viruses: (**A**). Left/up: Influenza virus (A/Alpaca/ID/2024(H5N1) had 7 NLGs, with a 68 NISS site (green bold); left/down: Influenza virus (A/polar-bear/ALK/2023) had 7 NLGs, with a 70 NNTN site (yellow bold) (**B**). Right/up: Influenza virus (A/Louisiana/2024(H5N1) had 7 NGLs with the substitution 50 NQSV (purple bold); right/down: Influenza virus (A/Calidris-alba/LIM(H5N1) had 7 predicted NGLs with 68 NISN (orange bold).

Conversely, 1 North American avian sample (peregrine-falcon/NY) had a change from 235 NGSC to 221 NNTL, with a notable reduction in the glycosylation potential from 0.73 to 0.47. In contrast, a decrease in the glycosylation potential was also observed in the avian sample (Northern-pintail/USA) and mammal samples (A/LOU/WT) to 0.67. Likewise, these 3 samples (Northern-pintail/USA and 2 A/LOU/WT) had a mutation detected in position 234 where V was replaced by I.

On the other hand, the sample (turkey/GER) had 6 NGLs, due to the loss of 88 NSSL glycosylation site predicted, and it matches with a substitution in the amino acid of 90 position from S to P. In addition, the isolate (chicken/JPN) showed a slight increase in the glycosylation potential at the 146 NGTV site of 0.68 to 0.7377, accompanied by a substitution in the amino acid position 150 from K to E.

4. Discussion

Noteworthy, the threat of Influenza H5N1 viruses globally within the One Health framework is undeniable, and one of the significant challenges of our era is to focus on limiting exposure and preventing the virus spread. Considering animal welfare, and the protection of threatened species is a breaking point, in order to maintain a sustainable connection between nature and people. Current facts such as the overexploitation of wild and domestic animals, unsustainable production systems [33], as well as poor water quality, with high loads of ubiquitous environmental pollutants, such as heavy metals (with immunosuppressive effects) [34,35], strongly impact aquatic habitats, altering water/sediment quality, and it directly affects microorganisms and birds/marine mammals, the principal reservoirs, and victims of Influenza viruses. Therefore, thinking about ecology facts and holistic perspectives to mitigate the transmission risk (including zoonosis reverse) between mammals and humans is so crucial. Likewise, the understanding of Influenza evolution plays a vital role in the mutations analysis and enhances viral fitness over time, and what better example than the amino acid change HA:190D in the pandemic Influenza virus of the 1918 H1N1 that switched its binding preference from SAα2,3Gal to SAα2,6Gal [36].

In the extended listing of amino acid variations in the present study, it was observed that some of them have already been described or studied, but the majority of recent mutations found still have unknown biological functions. Certain variations were detected in both bird and mammal species highlighted in spite of these having been classified previously as a warning level 1 [8] or having been detected with a different amino acid combination have been considered of importance: HA:11L, HA:52A, HA:104G, HA:152S, HA:211I, HA:336N/R, HA:492D, HA:527I, NA:6R, NA:10T, NA:70N, NA:405T, NA:234I, NA:329S, NA:436V, NP:377N, PA:354F, PA:545V, PA:614D/S, PA-X:250Q, PB2:9N, NS1:67G/Q, NP:119V/T, PA:322V/L, PB2:532L, PB2:539V, PB2:666I, PB2:670R, PB1:211K, PB1:384P/T/A, PB1:621K, PB1:738G, PB1:16D, PB1:154S, PB1:172D, PB1:207R, PB1:215K, PB1:264D, PB1:375N, PB1:548F, PB1:694S, PB1-F2:4G, PB1-F2:7I/T/M, PB1-F2:8Q, PB1-F2:12S PB1-F2:17S, PB1-F2:18T, PB1-F2:20R, PB1-F2:21R, PB1-F2:22E, PB1:36T, PB1-F2:40G, PB1-F2:42Y, PB1-F2:49A, PB1-F2:56A, PB1-F2:57C/F, PB1-F2:58W, PB1-F2:65R, PB1-F2:75L, PB1-F2:82S, PB1-F2:84S, PB1-F2:90N, PA-X:193S, PA-X:245N. Moreover, mutations found in only 1 animal (mammal): HA:226T [37], HA:310V; HA:520R, HA:432R, PA:336M, PA:351G, PA:399V, PA:404S, PA:459V, PA:486M, PA:489S, PA:538G, PA:655F, PA:655F, PB2:79G, PB2:152V, PB2:190R, PB1:251K, PB2:190R, PB2:251K, PB2:255A, PB2:274V, PB2:346A, PB2:353R, PB2:596A, PB2:663R, PB2:667I, PB2:670R, PB2:677K, PB2:715S, PB1:40I, PB1:171A, PB1:291A, PB1:390G, PB1:533S, PB1:660I, may also be considered for future analysis based on the limited information associated. In general, a greater number of mutations were detected in North America isolates, particularly in the most recent ones coming from mammals.

Likewise, a striking fact was observed in certain mammal isolates: the presentation of the same amino acid variation in both PA and PA-X proteins simultaneously, such as case of mutations PA:13V, PA-X:13V, PA:36T PA-X:36T, PA:45S, PA-X:45S, PA:70V, PA-X:70V, PA:75Q, PA-X:75Q, PA:142E, PA-X:142E, PA:160E, PA-X:160E, PA:211I, PA-X:211Y. In addition, the following group mutations are noteworthy and warrant further research considering the key role of the PA protein in the emergence of pandemic viruses: PA20T, PA-X:20T, PA:57Q, PA-X:57Q, PA:85T, PA-X:85T, PA:86I, PA-X:86I, (South America mammals/birds) and PA:61I, PA-X:61I, PA:68S, PA-X:68S, PA:100I, PA-X:100I (North America mammals/birds). The 2009 H1N1 (H1N1pdm09) virus gained multiple mutations in the PA gene through activation polymerase activity in mammalian cells, even in the absence of previously identified host adaptive mutations in others polymerase genes. In the same way, the activity of the accessory protein PA-X varies between different hosts and could play a role in host adaptation [38]. Now, studies of H5N1 viruses have found that mutations PA:343T and PA:383D could activate polymerase activity in human cells in the "absence" of key mutations in PB2, such as E627K or D701N [39,40]. Additionally, considering the most recent mammal isolates of H5N1 viruses have accumulated too many new mutations mainly in NA and PB1-F2 proteins (NA:23V, NA:44N, NA:45H, NA:48T, NA:53V, NA:62I, NA:67I, NA:74L, NA:75I, NA:81D, NA:82P, NA:84A, NA:221S, NA:241I, NA:257R, NA:286S, NA:288V, NA:329S, PB1-F2:4G, PB1-F2:7I/T/M, PB1-F2:8Q, PB1-F2:12S, PB1-

F2:18T, PB1-F2:20R, PB1-F2:21R, PB1-F2:22E, PB1:36T, PB1-F2:40G, PB1-F2:42Y, PB1-F2:49A, PB1-F2:58W, PB1-F2:65R, PB1-F2:75L, PB1-F2:82S, PB1-F2:84S), as well as PA:201V, PA:322V, PA:348L, PA:626R, PA-X:195K), it is recommended to analyze these variations carefully, in particular PB1-F2 due to the high number of changes found in a protein of only 87–90 amino acids, for which data from the reference genome (A/goose/Guandong/1996(H5N1) is not available. A study carried out in highly pathogenic H5N1 avian influenza viruses and the H1N1 pandemic strain of 1918 found that a single point mutation from N to S at position 66 of the PB1-F2 protein dramatically increased the virulence, and identified the antagonism to interferon [41,42]. Also, it is well-known that PB1-F2 had proapoptotic functions in immune cells and a pathogenicity potential increased through of dysregulation of the inflammatory response [43].

In addition, some key mutations associated with changes in glycosylation patterns have previously been observed such as HA:304N, NA:70N, NA:90P, NA:237F, NA:364N [44], and NA:329S [45]. Our specimens analyzed had included to the list: HA:211A, HA:499R, HA:102T, NA:67I, NA:234I, NA:150E and NA:50V. Its variations were found between mammals and non-mammals' species, hence future analysis is required. Glycosylation process is an indispensable factor for the infectivity, survival and transmissibility of the Influenza virions and NLG are an essential component in the adaptation of Influenza viruses to new hosts [23]. In the case of HA protein, the addition of oligosaccharides has been associated with changes in the ability to bind to cellular receptors, interaction with neutralizing antibodies (immune evasion), properly protein fold, fusion process, efficient transport, stability, fit budding, and virulence [22,46]; nonetheless, the exact function acquired through of a glycosylation is related with the "specific" area of HA is being strengthened: HA stem region is paramount for the membrane fusion, and correct protein folding; while the head region is related to mislead the immune system (masking antigenic sites of the receptor binding domain) [32], as well as the receptor-binding site (RBS), that influence viral receptor binding preferences [47]. Studies of the H1N1 virus have shown first that different antibody responses when HA protein was glycosylated [48] and second the appearance of a new glycosylation site (179) in HA protein, observed exclusively during the specific evolutionary phases of the seasonal virus strain, and it was suggested that the association of the emergence of this new glycosylation site with the increased incidence of Influenza A cases in 2023 [49]. On the other hand, the functions of glycosylation's in NA protein remains still unclear, there is relatively little research on this protein [22,50], but due to the opposing roles of HA and NA, it is recommended that both proteins should be studied jointly: the stability between HA binding and NA cleavage action is "essential" for overcoming host barriers and adaptation to new host species; so any important change observed in glycosylation patterns of any of these proteins, should be considered in detail since it could be a predicting feature of future pandemic/panzootia of Influenza viruses [47]. Our understanding of how glycosylation affects viral fitness is still limited; thus, it is recommended to carry out complementary analysis such as protein modeling and mass spectrophotometry based on comparative proteomics [51] to discern more about all the NLGs differences found in H5N1 viruses, especially when new glycosylation forms might appear and occur suddenly in a near future [32].

Otherwise, an important point to stand out under One Health approach is the finding of mutations associated with possible mild/strong antiviral resistance such as case of amantadine in mutation M2:27A/I (2 mammals, 1 bird) [52], as well as, NA:436V (1 mammal, 1 bird) [53] and NA:432R (1 mammal) [54]. Although, they have only been detected in a few animals, this fact is particularly important because there are mutations related to antiviral resistance to amantadine or oseltamivir. Furthermore, other variations detected, NA:103D, NA:106I, NA:107R, NA:141N/D, NA:142D, NA:144H, NA:153S, NA:188I/F,

NA:248N, NA:354G, NA:400S, NA429G, NA:438T, have shown different sensitive grades to oseltamivir, zanamivir, peramivir, and laninamivir [29]. The antiviral resistance has been more widely studied in H1N1 and H3N2 viruses, reporting the sporadic emergence of drug-resistant strains through compensatory mutations, underscoring the need to strengthen global surveillance and the urgency of new antiviral strategies to different strains of Influenza viruses [55,56]. A recent study of H5N1 viruses found a reassortant virus derived from a lineage of low pathogenicity avian Influenza virus, circulating in poultry with NA:275H, a marker of resistance to the oseltamivir. It is well known that antiviral resistance mutations favor high fatality cases and may necessitate a reevaluation of Influenza treatment strategies [57].

Finally, in general terms, we found some interesting facts related to Influenza virus mutational studies: (1) limited global database to keep up to date mutational and glycosylation profiles, in both the ancient Influenza viruses as the emerging/reemerging subtypes (not only H5N1). The bioinformatics servers upload all the information available, but are still required to have more analytical tools and data, for example, about the highly variable influenza accessory proteins in different numbering; (2) Influenza mutations have different nomenclature (numbering H1, H3, H5...), which makes it difficult to follow-up: for instance the mutation HA:S149A (H5 numbering) or S137A in (H3 numbering) [58,59], assigned warning level 3, (the most significant) [8], has been detected only by one of the tools used (in any numbering). So, it is a classic example of how mutational information of Influenza viruses does not have a consensus at all; (3) not all bioinformatics tools offer the option of selecting the "specific" virus variant with which you want to compare a mutation, and it is essential. The results differ greatly, for example, if you compare with a different variant, a vaccine candidate, a reference genome or one of the most recent H5N1 isolates now circulating. For instance, the mutation NA:329S has been recently related with changes in glycosylation patterns in viruses H3N2 [45], and there may be uncertainty, if it is similar to viruses of different numbering such as H5N1, H7N7, H1N1, H9N2; (4) a lack of data to reference genome (A/goose/Guandong/1996(H5N1) of accessory proteins PA-X and PB1-F2. As has been shown in the findings, there are many key mutations in these two proteins, which warrant further research.

5. Conclusions

Our study identified several key mutations between the 55 highly pathogenic avian Influenza A viruses (HPAIV) H5N1 isolates in outbreaks that occurred in 2022 and 2024, in the Americas region at a multi segment level such as HA:211I, NA:10T, NP:377N, PA:322V/L, PB2:539V, PB1:207R, PA-X:250Q, and PB1-F2:42Y, among others for which there is still little evidence of their possible biological function; it emphasizes the importance of carrying out meticulous surveillance in humans and animals to track crucial mutations. Additionally, the most relevant finding was the identification of a patron of cognate mutations: PA:13V, PA-X:13V, PA20T, PA-X:20T, PA:36T, PA-X:36T, PA:45S, PA-X:45S, PA:57Q, PA-X:57Q, PA:68S, PA-X:68S, PA:70V, PA-X:70V, PA:75Q, PA-X:75Q, PA:85T, PA-X:85T, PA:86I, PA-X:86I, PA:100I, PA-X:100I, PA:61I, PA-X:61I, PA:142E, PA-X:142E, PA:160E, PA-X:160E, PA:211I, PA-X:211Y, which highlight the role of the PA protein in the evolution of Influenza viruses through the activation of polymerase activity in the mammalian host, even in the absence of recognized host adaptive mutations in other polymerases such as PB2 [39]. In addition, the scanning of glycosylation profiles of H5N1 viruses highlights the loss/acquisition of NLG active sites 209NNTN, 100 NPTT, 302NSSM in HA protein and 70NNTN, 68NISS, 50NGSV in NA, providing valuable insights into viral evolution, notwithstanding, the limitations of the own study, and that the knowledge in this subject is still limited. The emergence of extra mutations and new glycosylation forms detected in H5N1 viruses underscore the necessity for further analysis and fosters science-based consensus to enhance our understanding of the evolution of these changeable viruses. Thus, reinforcing public databases is recommended, as is updating the H5N1 virus mutational profiles and including NLGs relevant sites.

Supplementary Materials: The following supporting information can be downloaded at: https://www.mdpi.com/article/10.3390/vetsci12040392/s1, S1: Listing of data sets/mutations found: S1.Table S1. Listing of H5N1 Influenza A viruses data set (NCBI) used in the study; S1.Table S2. Listing of mutations found in H5N1 Influenza viruses and its characteristics; S1.Table S3. Unique mutations (only detected in 1–3 samples) in H5N1 viruses identified per each protein segment; S1. Table S4: Mutations identified in H5N1 viruses per samples group (4 or more isolates); S1.Table S5. Mutations detected in all H5N1 viruses analyzed; S1.Table S6. Mutations found in almost all H5N1 viruses with some exceptions. S2: Complete mutational data analysis; S3: FluMut data analysis; S4: Genoflu data analysis S5: Listing of N-Linked Glycosylations (NLGs) profiles.

Author Contributions: Conceptualization, methodology, S.L.-C.; software, D.S.-A.; validation, S.L.-C.; formal analysis, S.L.-C., L.A.-A. and D.S.-A.; investigation, S.L.-C.; resources, S.L.-C.; data curation, S.L.-C., L.A.-A. and D.S.-A.; writing—original draft preparation, S.L.-C.; writing—review and editing, E.M.-Z. and E.M.-H.; visualization, L.A.-A.; supervision, E.M.-Z.; project administration, S.L.-C. and E.M.-H. All authors have read and agreed to the published version of the manuscript.

Funding: This research received no external funding.

Institutional Review Board Statement: The study was conducted in accordance with the Declaration of Helsinki and approved by the Institutional Ethics Committee of Biological Sciences code N° 086-2024-CBE-FCB-UNMSM.

Informed Consent Statement: Not applicable.

Data Availability Statement: Data are contained within the article and Supplementary Material.

Acknowledgments: We thank the staff of Pantanos of Villa Wildlife Refuge SERNANPfor their distinguished contribution.

Conflicts of Interest: The authors declare no conflicts of interest.

References

1. Duriez, O.; Sassi, Y.; Le Gall-Ladevèze, C.; Giraud, L.; Straughan, R.; Dauverné, L.; Terras, A.; Boulinier, T.; Choquet, R.; Van De Wiele, A.; et al. Highly pathogenic avian influenza affects vultures' movements and breeding output. *Curr. Biol.* **2023**, *33*, 3766–3774.e3. [CrossRef] [PubMed]
2. Caliendo, V.; Martin, B.B.; Fouchier, R.A.M.; Verdaat, H.; Engelsma, M.; Beerens, N.; Slaterus, R. Highly Pathogenic Avian Influenza Contributes to the Population Decline of the Peregrine Falcon (*Falco peregrinus*) in The Netherlands. *Viruses* **2024**, *17*, 24. [CrossRef] [PubMed]
3. Giacinti, J.A.; Signore, A.V.; Jones, M.E.; Bourque, L.; Lair, S.; Jardine, C.; Stevens, B.; Bollinger, T.; Goldsmith, D.; British Columbia Wildlife AIV Surveillance Program (BC WASP); et al. Avian influenza viruses in wild birds in Canada following incursions of highly pathogenic H5N1 virus from Eurasia in 2021–2022. *mBio* **2024**, *15*, e0320323. [CrossRef] [PubMed]
4. Banyard, A.C.; Bennison, A.; Byrne, A.M.P.; Reid, S.M.; Lynton-Jenkins, J.G.; Mollett, B.; De Silva, D.; Peers-Dent, J.; Finlayson, K.; Hall, R.; et al. Detection and spread of high pathogenicity avian influenza virus H5N1 in the Antarctic Region. *Nat. Commun.* **2024**, *15*, 7433. [CrossRef]
5. Uhart, M.M.; Vanstreels, R.E.T.; Nelson, M.I.; Olivera, V.; Campagna, J.; Zavattieri, V.; Lemey, P.; Campagna, C.; Falabella, V.; Rimondi, A. Epidemiological data of an influenza A/H5N1 outbreak in elephant seals in Argentina indicates mammal-to-mammal transmission. *Nat. Commun.* **2024**, *15*, 9516. [CrossRef]
6. Tomás, G.; Marandino, A.; Panzera, Y.; Rodríguez, S.; Wallau, G.L.; Dezordi, F.Z.; Pérez, R.; Bassetti, L.; Negro, R.; Williman, J.; et al. Highly pathogenic avian influenza H5N1 virus infections in pinnipeds and seabirds in Uruguay: Implications for bird–mammal transmission in South America. *Virus Evol.* **2024**, *10*, veae031. [CrossRef]

7. Ulloa, M.; Fernández, A.; Ariyama, N.; Colom-Rivero, A.; Rivera, C.; Nuñez, P.; Sanhueza, P.; Johow, M.; Araya, H.; Torres, J.C.; et al. Mass mortality event in South American sea lions (*Otaria flavescens*) correlated to highly pathogenic avian influenza (HPAI) H5N1 outbreak in Chile. *Veter Q.* **2023**, *43*, 1–10. [CrossRef]
8. Oguzie, J.U.; Marushchak, L.V.; Shittu, I.; Lednicky, J.A.; Miller, A.L.; Hao, H.; Nelson, M.I.; Gray, G.C. Avian Influenza A(H5N1) Virus among Dairy Cattle, Texas, USA. *Emerg. Infect. Dis.* **2024**, *30*, 1425–1429. [CrossRef]
9. Nguyen, T.Q.; Hutter, C.; Markin, A.; Thomas, M.; Lantz, K.; Lea Killian, M.; Janzen, G.M.; Vijendran, S.; Wagle, S.; Inderski, B.; et al. Emergence and interstate spread of highly pathogenic avian influenza A(H5N1) in dairy cattle. *biorxiv* **2024**. [CrossRef]
10. Bordes, L.; Vreman, S.; Heutink, R.; Roose, M.; Venema, S.; Pritz-Verschuren, S.B.E.; Rijks, J.M.; Gonzales, J.L.; Germeraad, E.A.; Engelsma, M.; et al. Highly Pathogenic Avian Influenza H5N1 Virus Infections in Wild Red Foxes (Vulpes vulpes) Show Neurotropism and Adaptive Virus Mutations. *Microbiol. Spectr.* **2023**, *11*, e0286722. [CrossRef]
11. Barry, K.T.; Tate, M.D. Flu on the Brain: Identification of Highly Pathogenic Influenza in the Brains of Wild Carnivores in The Netherlands. *Pathogens* **2023**, *12*, 1111. [CrossRef] [PubMed]
12. Chothe, S.K.; Srinivas, S.; Misra, S.; Nallipogu, N.C.; Gilbride, E.; LaBella, L.; Mukherjee, S.; Gauthier, C.H.; Pecoraro, H.L.; Webb, B.T.; et al. Marked neurotropism and potential adaptation of H5N1 clade 2.3.4.4.b virus in naturally infected domestic cats. *Emerg. Microbes Infect.* **2024**, *14*, 2440498. [CrossRef] [PubMed]
13. Ly, H. Highly pathogenic avian influenza H5N1 virus infection of companion animals. *Virulence* **2023**, *15*, 2289780. [CrossRef] [PubMed]
14. Pulit-Penaloza, J.A.; Brock, N.; Belser, J.A.; Sun, X.; Pappas, C.; Kieran, T.J.; Thakur, P.B.; Zeng, H.; Cui, D.; Frederick, J.; et al. Highly pathogenic avian influenza A(H5N1) virus of clade 2.3.4.4b isolated from a human case in Chile causes fatal disease and transmits between co-housed ferrets. *Emerg. Microbes Infect.* **2024**, *13*, 2332667. [CrossRef]
15. Duarte, P.M.; El-Nakeep, S.; Shayestegan, F.; Tazerji, S.S.; Malik, Y.S.; Roncada, P.; Tilocca, B.; Gharieb, R.; Hogan, U.; Ahmadi, H.; et al. Addressing the recent transmission of H5N1 to new animal species and humans, warning of the risks and its relevance in One-Health. *Ger. J. Microbiol.* **2024**, *4*, 39–53. [CrossRef]
16. EFSA Panel on Animal Health and Animal Welfare (AHAW); ECDC; Alvarez, J.; Boklund, A.; Dippel, S.; Dórea, F.; Figuerola, J.; Herskin, M.S.; Michel, V.; Miranda Chueca, M.Á.; et al. Preparedness, prevention and control related to zoonotic avian influenza. *EFSA J.* **2025**, *23*, e9191. [CrossRef]
17. Dholakia, V.; Quantrill, J.L.; Richardson, S.; Pankaew, N.; Brown, M.D.; Yang, J.; Capelastegui, F.; Masonou, T.; Case, K.M.; Aejian, J.; et al. Polymerase mutations underlie early adaptation of H5N1 influenza virus to dairy cattle and other mammals. *bioRxiv* **2025**. [CrossRef]
18. Lin, T.-H.; Zhu, X.; Wang, S.; Zhang, D.; McBride, R.; Yu, W.; Babarinde, S.; Paulson, J.C.; Wilson, I.A. A single mutation in bovine influenza H5N1 hemagglutinin switches specificity to human receptors. *Science* **1979**, *386*, 1128–1134. [CrossRef]
19. Xiao, Y.; Sheng, Z.-M.; Williams, S.L.; Taubenberger, J.K. Two complete 1918 influenza A/H1N1 pandemic virus genomes characterized by next-generation sequencing using RNA isolated from formalin-fixed, paraffin-embedded autopsy lung tissue samples along with evidence of secondary bacterial co-infection. *mBio* **2024**, *15*, e0321823. [CrossRef]
20. Sealy, J.E.; Peacock, T.P.; Sadeyen, J.-R.; Chang, P.; Everest, H.J.; Bhat, S.; Iqbal, M. Adsorptive mutation and N-linked glycosylation modulate influenza virus antigenicity and fitness. *Emerg. Microbes Infect.* **2020**, *9*, 2622–2631. [CrossRef]
21. Sun, Y.; Zhu, Y.; Zhang, P.; Sheng, S.; Guan, Z.; Cong, Y. Hemagglutinin glycosylation pattern-specific effects: Implications for the fitness of H9.4.2.5-branched H9N2 avian influenza viruses. *Emerg. Microbes Infect.* **2024**, *13*, 2364736. [CrossRef] [PubMed]
22. Bao, D.; Xue, R.; Zhang, M.; Lu, C.; Ma, T.; Ren, C.; Zhang, T.; Yang, J.; Teng, Q.; Li, X.; et al. N-Linked Glycosylation Plays an Important Role in Budding of Neuraminidase Protein and Virulence of Influenza Viruses. *J. Virol.* **2021**, *95*, 10–1128. [CrossRef] [PubMed]
23. Kim, P.; Jang, Y.H.; Bin Kwon, S.; Lee, C.M.; Han, G.; Seong, B.L. Glycosylation of hemagglutinin and neuraminidase of influenza a virus as signature for ecological spillover and adaptation among influenza reservoirs. *Viruses* **2018**, *10*, 183. [CrossRef] [PubMed]
24. Fouchier, R.A.M.; Bestebroer, T.M.; Herfst, S.; Van Der Kemp, L.; Rimmelzwaan, G.F.; Osterhaus, A.D.M.E. Detection of influenza A viruses from different species by pcr amplification of conserved sequences in the matrix gene. *J. Clin. Microbiol.* **2000**, *38*, 4096–4101. [CrossRef]
25. WHO. WHO Information for the Molecular Detection of Influenza Viruses. 2017, pp. 1–60. Available online: https://cdn.who.int/media/docs/default-source/influenza/molecular-detention-of-influenza-viruses/protocols_influenza_virus_detection_feb_2021.pdf (accessed on 15 April 2025).
26. Landazabal-Castillo, S.; Suarez-Agüero, D.; Alva-Alvarez, L.; Mamani-Zapana, E.; Mayta-Huatuco, E. Highly pathogenic avian influenza A virus subtype H5N1 (clade 2.3.4.4b) isolated from a natural protected area in Peru. *Microbiol. Resour. Announc.* **2024**, *13*, e0041724. [CrossRef]
27. Madeira, F.; Madhusoodanan, N.; Lee, J.; Eusebi, A.; Niewielska, A.; Tivey, A.R.N.; Lopez, R.; Butcher, S. The EMBL-EBI Job Dispatcher sequence analysis tools framework in 2024. *Nucleic Acids Res.* **2024**, *52*, W521–W525. [CrossRef]
28. GNU Affero. FluMut. 2024. Available online: https://github.com/izsvenezie-virology/FluMut (accessed on 20 January 2025).

29. Philippine Genome Center. FluSurver in GISAID. 2024. Available online: https://gisaid.org/database-features/flusurver-mutations-app/ (accessed on 21 February 2024).
30. Youk, S.; Torchetti, M.K.; Lantz, K.; Lenoch, J.B.; Killian, M.L.; Leyson, C.; Bevins, S.N.; Dilione, K.; Ip, H.S.; Stallknecht, D.E.; et al. H5N1 highly pathogenic avian influenza clade 2.3.4.4b in wild and domestic birds: Introductions into the United States and reassortments, December 2021–April 2022. *Virology* **2023**, *587*, 109860. [CrossRef]
31. Gupta, R.; Brunak, S. Prediction of glycosylation across the human proteome and the correlation to protein function. *Pac. Symp. Biocomput.* **2002**, *7*, 310–322.
32. Boeijen, M. Glycosylation of the Influenza A Virus Hemagglutinin Protein. 2013. Available online: https://studenttheses.uu.nl/bitstream/handle/20.500.12932/14431/glycosylation%20of%20the%20influenza%20A%20virus%20hemagglutinin%20protein1.pdf?sequence=1 (accessed on 15 April 2025).
33. Lambertucci, S.A.; Santangeli, A.; Plaza, P.I. The threat of avian influenza H5N1 looms over global biodiversity. *Nat. Rev. Biodivers.* **2025**, *1*, 7–9. [CrossRef]
34. Ahamad, M.I.; Yao, Z.; Ren, L.; Zhang, C.; Li, T.; Lu, H.; Mehmood, M.S.; Rehman, A.; Adil, M.; Lu, S.; et al. Impact of heavy metals on aquatic life and human health: A case study of River Ravi Pakistan. *Front. Mar. Sci.* **2024**, *11*, 1374835. [CrossRef]
35. Zhang, H.; Wang, J.; Zhang, K.; Shi, J.; Gao, Y.; Zheng, J.; He, J.; Zhang, J.; Song, Y.; Zhang, R.; et al. Association between heavy metals exposure and persistent infections: The mediating role of immune function. *Front. Public. Health* **2024**, *12*, 1367644. [CrossRef] [PubMed]
36. Glaser, L.; Stevens, J.; Zamarin, D.; Wilson, I.A.; García-Sastre, A.; Tumpey, T.M.; Basler, C.F.; Taubenberger, J.K.; Palese, P. A Single Amino Acid Substitution in 1918 Influenza Virus Hemagglutinin Changes Receptor Binding Specificity. *J. Virol.* **2005**, *79*, 11533–11536. [CrossRef] [PubMed]
37. Watanabe, Y.; Ibrahim, M.S.; Ellakany, H.F.; Kawashita, N.; Mizuike, R.; Hiramatsu, H.; Sriwilaijaroen, N.; Takagi, T.; Suzuki, Y.; Ikuta, K. Acquisition of Human-Type Receptor Binding Specificity by New H5N1 Influenza Virus Sublineages during Their Emergence in Birds in Egypt. *PLOS Pathog.* **2011**, *7*, e1002068. [CrossRef] [PubMed]
38. Lutz, M.M.; Dunagan, M.M.; Kurebayashi, Y.; Takimoto, T. Key role of the influenza a virus pa gene segment in the emergence of pandemic viruses. *Viruses* **2020**, *12*, 365. [CrossRef]
39. Arai, Y.; Kawashita, N.; Hotta, K.; Hoang, P.V.M.; Nguyen, H.L.K.; Nguyen, T.C.; Vuong, C.D.; Le, T.T.; Le, M.T.Q.; Soda, K.; et al. Multiple polymerase gene mutations for human adaptation occurring in Asian H5N1 influenza virus clinical isolates. *Sci. Rep.* **2018**, *8*, 13066. [CrossRef]
40. Song, J.; Xu, J.; Shi, J.; Li, Y.; Chen, H. Synergistic Effect of S224P and N383D Substitutions in the PA of H5N1 Avian Influenza Virus Contributes to Mammalian Adaptation. *Sci. Rep.* **2015**, *5*, 10510. [CrossRef]
41. Conenello, G.M.; Zamarin, D.; Perrone, L.A.; Tumpey, T.; Palese, P. A single mutation in the PB1-F2 of H5N1 (HK/97) and 1918 influenza a viruses contributes to increased virulence. *PLOS Pathog.* **2007**, *3*, e141. [CrossRef]
42. Varga, Z.T.; Palese, P. The influenza A virus protein PB1-F2. *Virulence* **2011**, *2*, 542–546. [CrossRef]
43. Chakrabarti, A.K.; Pasricha, G. An insight into the PB1F2 protein and its multifunctional role in enhancing the pathogenicity of the influenza A viruses. *Virology* **2013**, *440*, 97–104. [CrossRef]
44. Roubidoux, E.K.; Sano, K.; McMahon, M.; Carreño, J.M.; Capuano, C.; Jiang, K.; Simon, V.; van Bakel, H.; Wilson, P.; Krammer, F. Novel Epitopes of the Influenza Virus N1 Neuraminidase Targeted by Human Monoclonal Antibodies. *J. Virol.* **2022**, *96*, e0033222. [CrossRef]
45. Anoma, S.; Bhattarakosol, P.; Kowitdamrong, E. Characteristics and evolution of hemagglutinin and neuraminidase genes of Influenza A(H3N2) viruses in Thailand during 2015 to 2018. *PeerJ* **2024**, *12*, e17523. [CrossRef] [PubMed]
46. Schulze, I.T. Effects of Glycosylation on the Properties and Functions of Influenza Virus Hemagglutinin. *J. Infect. Dis.* **1997**, *176*, S24–S28. [CrossRef] [PubMed]
47. Guo, X.; Zhou, Y.; Yan, H.; An, Q.; Liang, C.; Liu, L.; Qian, J. Molecular Markers and Mechanisms of Influenza A Virus Cross-Species Transmission and New Host Adaptation. *Viruses* **2024**, *16*, 883. [CrossRef] [PubMed]
48. Wang, C.-C.; Chen, J.-R.; Tseng, Y.-C.; Hsu, C.-H.; Hung, Y.-F.; Chen, S.-W.; Chen, C.-M.; Khoo, K.-H.; Cheng, T.-J.; Cheng, Y.-S.E.; et al. Glycans on influenza hemagglutinin affect receptor binding and immune response. *Proc. Natl. Acad. Sci. USA* **2009**, *106*, 18137–18142. [CrossRef]
49. Wang, N.; Lu, W.; Yan, L.; Liu, M.; Che, F.; Wang, Y.; Yang, C.; Lv, M.; Cheng, J.; Sun, Q.; et al. Epidemiological and genetic characterization of the influenza A (H1N1) virus in Hangzhou City in 2023. *Front. Public. Health* **2024**, *12*, 1464435. [CrossRef]
50. Chen, W.; Ma, T.; Liu, S.; Zhong, Y.; Yu, H.; Shu, J.; Wang, X.; Li, Z. *N*-Glycan Profiles of Neuraminidase from Avian Influenza Viruses. *Viruses* **2024**, *16*, 190. [CrossRef]
51. She, Y.-M.; Farnsworth, A.; Li, X.; Cyr, T.D. Topological N-glycosylation and site-specific N-glycan sulfation of influenza proteins in the highly expressed H1N1 candidate vaccines. *Sci. Rep.* **2017**, *7*, 10232. [CrossRef]
52. Ilyushina, N.A.; Govorkova, E.A.; Webster, R.G. Detection of amantadine-resistant variants among avian influenza viruses isolated in North America and Asia. *Virology* **2005**, *341*, 102–106. [CrossRef]

53. Kwon, J.J.; Choi, W.-S.; Jeong, J.H.; Kim, E.-H.; Lee, O.-J.; Yoon, S.-W.; Hwang, J.; Webby, R.J.; Govorkova, E.A.; Choi, Y.K.; et al. An I436N substitution confers resistance of influenza A(H1N1)pdm09 viruses to multiple neuraminidase inhibitors without affecting viral fitness. *J. Gen. Virol.* **2018**, *99*, 292–302. [CrossRef]
54. Nuss, J.M.; Whitaker, P.B.; Air, G.M. Identification of critical contact residues in the NC41 epitope of a subtype N9 influenza virus neuraminidase. *Proteins Struct. Funct. Bioinform.* **1993**, *15*, 121–132. [CrossRef]
55. Xu, J.; Luo, Q.; Huang, Y.; Li, J.; Ye, W.; Yan, R.; Zhou, X.; He, Z.; Liu, G.; Zhu, Q. Influenza neuraminidase mutations and resistance to neuraminidase inhibitors. *Emerg. Microbes Infect.* **2024**, *13*, 2429627. [CrossRef] [PubMed]
56. Rabie-Rudsari, M.; Behboudi, E.; Ranjkesh, A.; Kaveh, K.; Razavi-Nikoo, H.; Haghshenas, M.R.; Moradi, A. Molecular identification of neuraminidase gene mutations in influenza A/H1N1 and A/H3N2 isolates of Mazandaran province, north of Iran. *J. Glob. Antimicrob. Resist.* **2023**, *36*, 466–472. [CrossRef] [PubMed]
57. Signore, A.V.; Joseph, T.; Ranadheera, C.; Erdelyan, C.N.G.; Alkie, T.N.; Raj, S.; Pama, L.; Ayilara, I.; Hisanaga, T.; Lung, O.; et al. Neuraminidase reassortment and oseltamivir resistance in clade 2.3.4.4b A(H5N1) viruses circulating among Canadian poultry, 2024. *Emerg. Microbes Infect.* **2025**, *14*, 2469643. [CrossRef] [PubMed]
58. Kadi, H. A Comprehensive Analysis of H5N1 Evolution: Phylogenetic Insights and Emerging Mutations in Turkey's Avian Influenza Landscape. *Preprint* **2024**. [CrossRef]
59. Yang, C.-R.; King, C.-C.; Liu, L.-Y.D.; Ku, C.-C. FluConvert and IniFlu: A suite of integrated software to identify novel signatures of emerging influenza viruses with increasing risk. *BMC Bioinform.* **2020**, *21*, 316. [CrossRef]

veterinary sciences

Article

Scalable Production of Recombinant Adeno-Associated Virus Vectors Expressing Soluble Viral Receptors for Broad-Spectrum Inhibition of Porcine Reproductive and Respiratory Syndrome Virus Type 2

Xiaoming Liu [1,2,*,†], Nuo Xu [1,†], Xiaoli Song [3], Linlin Zhuang [1], Qiuping Shen [1] and Huaichang Sun [2,*]

1 The Department of Animal Husbandry and Veterinary Medicine, Jiangsu Vocational College of Agriculture and Forestry, Jurong 212400, China; nuoxu@jsafc.edu.cn (N.X.); zhuanglinlin@jsafc.edu.cn (L.Z.); shenqiuping@jsafc.edu.cn (Q.S.)
2 The College of Veterinary Medicine, Yangzhou University, Yangzhou 225009, China
3 Jiangsu Provincial Animal Disease Control Center, 124 Caochangmen Street, Nanjing 210036, China; sxl_gadpcc@126.com
* Correspondence: liuxiaoming@jsafc.edu.cn (X.L.); sunh@yzu.edu.cn (H.S.)
† These authors contributed equally to this work.

Simple Summary: We developed a scalable insect cell–baculovirus system to produce rAAV vectors expressing dual soluble viral receptors (Sn4D-Fc/SRCR59-Fc) against PRRSV. By systematically optimizing baculovirus co-infection ratio (0.5:1.0), initial cell density (2.0×10^6 cells/mL), fetal bovine serum concentration (2%), and temperature (30 °C), we achieved a scalable production of rAAV-SRCR59-Fc/Sn4D-Fc vectors with a titer of $5.4 \pm 0.9 \times 10^9$ infectious viral particles (IVPs)/mL in a 2 L bioreactor, representing a 170-fold increase compared with conventional flask-based methods. In vitro tests showed these SVRs reduced viral titers of diverse PRRSV strains by ~4.3 log, demonstrating broad-spectrum antiviral activity. This platform offers a durable, scalable solution to combat PRRSV, addressing vaccine limitations for swine health management.

Abstract: Porcine reproductive and respiratory syndrome virus (PRRSV) continues to be a major threat to the global swine industry, causing significant economic losses. To address this, we developed a scalable recombinant adeno-associated virus (rAAV)-based strategy for the delivery of soluble viral receptors (SVRs) to treat and potentially eliminate PRRSV infections. This strategy involves fusing the virus-binding domains of two key cellular receptors, sialoadhesin (Sn4D) and CD163 (SRCR5-9), with an Fc fragment. We then used an insect cell–baculovirus expression vector system to produce the rAAV-SRCR59-Fc/Sn4D-Fc vector. Through a series of optimizations, we determined the best conditions for rAAV production, including a baculovirus co-infection ratio of 0.5:1.0, an initial insect cell density of 2.0×10^6 cells/mL, a fetal bovine serum concentration of 2%, and a culture temperature of 30 °C. Under these optimized conditions, we achieved a high titer of rAAV-SRCR59-Fc/Sn4D-Fc in a 2 L bioreactor, reaching $5.4 \pm 0.9 \times 10^9$ infectious viral particles (IVPs)/mL. Notably, in vitro neutralization assays using a Transwell co-culture system demonstrated a 4.3 log reduction in viral titers across genetically diverse PRRSV-2 strains, including VR2332, JXA1, JS07, and SH1705. Collectively, this study provides a robust platform for large-scale rAAV production and highlights the potential of SVR-based gene therapy to address the antigenic diversity of PRRSV-2.

Keywords: PRRSV; soluble viral receptor fusions; adeno-associated virus vector; insect cell bioreactor; broad-spectrum antiviral activity

1. Introduction

Porcine reproductive and respiratory syndrome (PRRS) is a highly contagious swine infectious disease caused by porcine reproductive and respiratory syndrome virus (PRRSV), which is characterized by late-term maternal reproductive failure, increased mortality of newborn piglets, and respiratory disease in growing pigs [1,2]. PRRSV is classified into two major genotypes: European (Type 1) and North American (Type 2), which exhibit ~60% nucleotide identity [3,4]. Currently, the North American genotype remains the dominant circulating strain in China. At present, PRRS is mainly controlled by immunization with attenuated live vaccines [5]. Although the vaccine could induce relative immune protection against homologous strain infection, the immune protection against heterologous strains is debatable due to the risk of virulence conversion and virion spread [6,7]. In addition, PRRSV is a highly variable RNA virus with complex infection and immune evasion mechanisms [8,9]. Therefore, the improvement of new PRRS vaccines faces serious challenges, and a new anti-PRRSV strategy is urgently needed.

Porcine alveolar macrophages (PAMs) serve as the primary target cells for PRRSV-2 infection, with sialoadhesin (Sn) and CD163 acting as critical entry receptors [10,11]. Sn, a PAM-restricted immunoglobulin superfamily receptor containing 17 Ig-like domains, mediates viral attachment via its N-terminal Sn4D domains in a sialic acid-dependent manner, interacting specifically with the M/GP5 glycoprotein complex of PRRSV-2 [12]. CD163, a PAM-specific scavenger receptor cysteine-rich (SRCR) superfamily protein, has been identified as the cellular receptor for PRRSV-2 glycoprotein 2/4 (GP2/GP4). This interaction occurs through its fifth SRCR domain to facilitate viral entry, while domains 6–9 (SRCR59) are essential for maintaining structural integrity and enabling efficient virus-receptor interaction [11]. These findings highlight the potential of soluble viral-binding domains Sn4D and SRCR5-9 as novel antiviral agents. Previous studies using recombinant adenovirus vectors demonstrated that these soluble viral receptors (SVRs) effectively blocked PRRSV-2 infection both in vitro and in vivo; however, their therapeutic utility was limited by short-term transgene expression [13,14]. To address this, we recently developed a recombinant adeno-associated virus (rAAV) vector expressing dual SVRs (Sn4D-SRCR59-Fc and SRCR59-Fc/Sn4D-Fc) fused with porcine IgG1 Fc, achieving prolonged transgene expression in murine models [15].

rAAV is generally considered a promising gene transfer vector with low immunogenicity and long-term expression in animal models or clinical therapy [16,17]. Some investigators have demonstrated that some rAAV serotypes show high efficiency for gene transfer in specific tissues [18,19]. However, the classic low-titer production mode of rAAV cannot meet the requirements for large animal studies or human clinical trials. The insect cell–baculovirus expression vector system has been discovered as a reproducible, facile, and adaptable process to produce rAAV, with high biological interest [20,21]. In this study, we developed an optimized insect cell–baculovirus system to achieve scalable production of rAAV-SRCR59-Fc/Sn4D-Fc vectors, yielding high titers of $5.4 \pm 0.9 \times 10^9$ infectious viral particles (IVPs)/mL in a 2 L bioreactor. In vitro assays demonstrated that these vectors potently blocked infection by diverse PRRSV-2 strains (VR2332, JXA1, JS07, and SH1705) with a ~4.3 log reduction in viral titers, highlighting their broad-spectrum antiviral potential.

2. Materials and Methods

2.1. Cells

Spodoptera frugiperda (Sf9) cells (Invitrogen, Carlsbad, CA, USA) were seeded at 27 °C in ExpiSf™ CD medium (Gibco, Billings, MT, USA) augmented with 1% penicillin/streptomycin. The monkey kidney cell line Marc-145 (ATCC, Manassas, VA, USA) and the human embryonic kidney cell line HEK-293A (Agilent Technologies, Santa Clara, CA, USA) were seeded in Dulbecco's modified Eagle medium (DMEM, Gibco) containing 10% fetal bovine serum (FBS). Primary PAM cells were isolated from Landrace pigs [22], and the SV40-transformed PAM cell line 3D4/21 (ATCC, Manassas, VA, USA) was cultured in RPMI 1640 medium supplemented with 10% FBS, 2 mM L-glutamine, 1% non-essential amino acids, and 1 mM sodium pyruvate (Invitrogen, Carlsbad, CA, USA).

2.2. Virus

Recombinant Bac-RC (encoding AAV-2 rep and cap genes) and recombinant Bac-SRCR59-Fc/Sn4D-Fc (encoding the SRCR59-Fc/Sn4D-Fc transgene) were propagated and titrated in Sf9 cells [15]. PRRSV strain VR2332 (ATCC VR2332) is a prototype strain of genotype 2 [13]. PRRSV strains JXA1 [23], JS07 [24], and SH1705 [14] are highly pathogenic strains isolated in China. The four PRRSV-2 strains were propagated and titrated on MARC-145 cells.

2.3. Optimization of the Conditions for rAAV Production

rAAV-SRCR59-Fc/Sn4D-Fc was generated via co-infection of Sf9 cells with two recombinant baculoviruses: rBac-SRCR59-Fc/Sn4D-Fc and rBac-RC. Prior to optimization, rBac-RC and rBac-SRCR59-Fc/Sn4D-Fc were titrated on Sf9 cells by real-time fluorescence quantitative PCR (RTFQ PCR) using TB Green® Premix ExTaq™ II Kit (Invitrogen) and the primer pairs listed in Table 1. For optimization, Sf9 cells were cultured in 125 mL shake flasks (CORNING, Corning, NY, USA) with 20 mL medium. Initially, cells (1.0×10^6 cells/mL) were co-infected with rBac-SRCR59-Fc/Sn4D-Fc at different multiplicities of infection (MOI: 0.1, 0.5, 1.0, 5.0, or 10) and rBac-RC (MOI 1.0) to evaluate the effect of baculovirus concentration ratios on rAAV production. Subsequent experiments varied the initial insect cell density (0.5, 1.0, 2.0, or 3.0×10^6 cells/mL), fetal bovine serum concentration (0%, 2%, 5%, 10%), and incubation temperature (24 °C, 27 °C, 30 °C, or 33 °C). Viral titers were quantified using a standardized AAV-293 cell-based immunofluorescence assay with FITC-conjugated goat anti-pig IgG (KPL, Gaithersburg, MD, USA) at a 1:200 dilution, followed by automated image analysis (ImageXpress Micro XLS, Molecular Devices, San Jose, CA, USA) to count fluorescence-forming units (FFUs). Statistical significance was determined via two-tailed Student's *t*-test or Kruskal–Wallis test using SPSS 22.0.

Table 1. The PCR primers used in this study.

Gene	Primer Pair	Sequence (5′→3′)	Amplicon (bp)	Reference
AAV-Rep (of rBac-RC)	Sense	TGGTGGACGAGTGCTACA	113	[15]
	Antisense	CTCCGTGAGATTCAAACAGG		
P_{CMV} (of Bac-SRCR59-Fc/Sn4D-Fc)	Sense	ATGGTGATGCGGTTTTGGC	123	[15]
	Antisense	AGTCCCGTTGATTTTGGTG		

2.4. Preparation of Baculovirus-Infected Insect Cells (BIICs)

Under the optimized conditions, Sf9 cells were grown in 125 mL shake flasks with 20 mL of medium and co-infected with rBac-SRCR59-Fc/Sn4D-Fc and rBac-RC. When the cell diameter became larger and the cell viability remained >90%, the baculovirus-infected insect cells (BIICs) were recovered by low-speed centrifugation (300× *g*, for 10 min) and resuspended in serum-free cell cryopreservation culture medium (NCM Biotech, Suzhou, China). One-milliliter aliquots containing 1.0×10^7 cells were dispensed into cryotubes (CORNING, Corning, NY, USA) and placed at a temperature of −80 °C.

2.5. Large-Scale rAAV Production

The two-liter bioreactor (BioFlo 320, Eppendorf, Hamburg, Germany) was filled with the minimum working volume of insect serum-free growth medium. Sf9 cells (7.5×10^5/mL) were inoculated into the rocking platform bioreactor after the temperature was maintained at 27 °C, and 30% O_2 in an air mixture was provided as previously described [25]. Detect the cell number in a timely manner and perform fed-batch with fresh 2% FBS medium to maintain a cell density of 1.6~1.8 $\times 10^6$/mL [26]. Once the final working volume and a cell density of 2.0×10^6 cells/mL were reached, thawed BIICs aliquots (BIICs:Sf9 cells = 10^{-3}) were added to the culture with the temperature adjusted to 30 °C. Cell density or viability was measured using trypan blue dye exclusion and an automated cell counter (Countess™ 3 FL, Invitrogen, Carlsbad, CA, USA), and the same volume of cell medium was collected and rAAV was titrated every 24 h until t = 216 h post-infection (hpi). After the cell diameter became larger and the proportion of infected cells exceeded 90%, the cells were centrifuged for 15 min at 2500× *g* and suspended in lysis buffer (50 mM HEPES, 2 mM $MgCl_2$, pH 7.5) for rAAV purification.

2.6. Purification of Recombinant Adeno-Associated Viruses

After three cycles of freeze–thawing in lysis buffer, the cells were treated with benzonase nuclease (50 U/mL) and $MgSO_4$ (37.5 mM) and disrupted two times at 1200 bar using a High-Pressure Cell Disruptor (JNBIO, Guangzhou, China). The supernatants were extracted with 10% chloroform following centrifugation for 15 min at 4 °C at 10,000× *g* as previously described [15]. Next, the upper aqueous phases were incubated with 25 mM $CaCl_2$ for 1 h on ice. After 15 min of centrifugation at 10,000× *g* at 4 °C, the supernatants were incubated with 0.62 M NaCl and 8% PEG8000 for 3 h on ice as previously described [27]. Following 30 min of centrifugation at 3000× *g* at 4 °C, the purified rAAV was suspended in PBS buffer and titrated on AAV-293 cells.

2.7. Transmission Electron Microscopy

The purified rAAV was absorbed onto copper grids (400-mesh) at room temperature for 2.5 min and dried with filter paper. The grids were stained with 3% phosphotungstic acid for 2.5 min and observed under a transmission electron microscope (Tecnai 12, Philips, Eindhoven, Netherlands) at an acceleration voltage of 75 kV.

2.8. Immunofluorescence

AAV-293 or 3D4/21 cells were seeded onto 6-well plates and transduced individually with the rAAV-SRCR59-Fc/Sn4D-Fc. At 48 h post-transduction (HPT), the cells were fixed with 4% paraformaldehyde, permeabilized with 0.5% Triton X-100, and blocked in PBSM (5% defatted milk powder in PBS, pH 7.4). Immunofluorescence was performed as previously described [28] using anti-pig IgG.

2.9. Western Blotting

The purified rAAV was separated by 12% SDS–PAGE and transferred onto a nitrocellulose membrane (Merk, Darmstadt, Germany) using a Trans-Blot Turbo instrument (Bio-Rad, Hercules, CA, USA). After blocking with 5% skim milk powder in PBST (0.1% Tween 20 in PBS), Western blotting was performed using homemade mouse anti-AAV serum (1:200) as the first antibody and DyLight 800-labelled goat anti-mouse IgG (1:10,000; KPL, Gaithersburg, MD, USA) as the second antibody. The hybridization signals were scanned using an Infrared Imaging System (Odyssey DLX, LI–COR, Lincoln, NE, USA) at 800 nm as previously described [15]. The SVR fusions SRCR59-Fc and Sn4D-Fc were purified using Protein A Sefinose Columns (Sangon, Shanghai, China) following the manufacturer's instructions, as previously described [13]. The purified proteins (200 ng) were separated by 12% SDS–PAGE and analyzed by Western blotting using DyLight 800-labelled goat anti-pig IgG (1:10,000), as previously described [15].

2.10. Viral Infection Blocking Assay

To evaluate the anti-PRRSV activity of the two SVR fusion proteins (SRCR59-Fc and Sn4D-Fc) against diverse PRRSV strains (VR2332, JXA1, JS07, and SH1705), 3D4/21 cells were cultured in the upper chambers of 24-well Transwell plates (CORNING, Corning, NY, USA) and transduced with rAAV-SRCR59-Fc/Sn4D-Fc at a fixed MOI of 100, with empty AAV serving as the negative control. At 48 h post-transduction, PAM cells in the lower chambers were infected with the indicated PRRSV strains at an MOI of 0.5. After 24 h of co-cultivation, PAM cells were lysed by three cycles of freeze–thawing (−80 °C for 10 min/37 °C for 5 min) to release viral particles for PRRSV titration on MARC-145 cells using the 50% tissue culture infective dose (TCID50) assay as previously described [29]. All experiments were performed in triplicate, and data were analyzed using Student's *t*-test.

2.11. Statistical Analysis

Statistical analysis was performed using SPSS Statistics 22. The results were considered to be statistically significant at $p < 0.05$ or extremely significant at $p < 0.01$. Three independent assays were performed for each separate data point, and the results are presented as the mean ± standard deviation (SD).

3. Results

3.1. Determination of the Optimal Packaging Conditions for rAAV

Considering that an excessively high viral titer can induce rapid cell death and subsequently compromise the packaging efficiency of rAAV, this study primarily investigated the influence of the baculovirus co-infection ratio on AAV production. Sf9 cells were seeded at an initial density of 1.0×10^6 cells/mL into 125 mL shake flasks filled with 20 mL of medium supplemented with 10% FBS. The cultures were maintained at 27 °C with a shaking speed of 100 rpm. Subsequently, Sf9 cells were co-infected with rBac-SRCR59-Fc/Sn4D-Fc at different multiplicities of infection (MOIs: 0.1, 0.5, 1.0, 5.0, or 10) and rBac-RC at an MOI of 1.0. After 72 h of co-infection, rAAVs were collected and titrated on AAV-293 cells by immunofluorescence using FITC-labeled goat anti-pig IgG and expressed as FFUs/mL. The results demonstrated that when the co-infection was carried out at an MOI ratio of 0.5:1.0 (rBac-SRCR59-Fc/Sn4D-Fc:rBac-RC), the rAAV achieved the highest titer of $4.0 \pm 0.5 \times 10^6$ IVPs/mL. This titer was highly significantly greater than those at MOI ratios of 5.0:1.0 ($1.7 \pm 0.4 \times 10^6$ IVPs/mL) and 10:1.0 ($0.6 \pm 0.2 \times 10^6$ IVPs/mL). Moreover, it was significantly higher than those at 1.0:1.0 ($2.4 \pm 0.1 \times 10^6$ IVPs/mL) and 0.1:1.0 ($2.6 \pm 0.1 \times 10^6$ IVPs/mL).

Testing initial insect cell densities demonstrated that the rAAV packaging yield was maximized at densities of 2.0×10^6 cells/mL ($6.1 \pm 0.1 \times 10^6$ IVPs/mL), which was highly significantly higher than yields observed at densities of 0.5×10^6 cells/mL ($0.5 \pm 0.1 \times 10^6$ IVPs/mL) or 3.0×10^6 cells/mL ($2.1 \pm 0.3 \times 10^6$ IVPs/mL), and significantly higher than densities of 1.0×10^6 cells/mL ($4.0 \pm 0.1 \times 10^6$ IVPs/mL) (Figure 1B). The concentration of FBS in insect cell growth medium also affected the production of rAAV. The highest rAAV packaging yields was observed in insect media containing 2% ($6.1 \pm 0.1 \times 10^6$ IVPs/mL), 5% ($6.2 \pm 0.3 \times 10^6$ IVPs/mL), or 10% ($6.0 \pm 0.3 \times 10^6$ IVPs/mL) FBS, compared to serum-free medium ($2.5 \pm 0.1 \times 10^6$ IVPs/mL) (Figure 1C). In addition, when considering the culture temperature, the number of rAAVs at 30 °C ($1.9 \pm 0.3 \times 10^7$ IVPs/mL) was highly significantly higher than that at 27 °C (6.1 ± 0.1 × 10^6 IVPs/mL) or 24 °C (1.0 ± 0.1 × 10^6 IVPs/mL), and no viral infectivity was detected at 33 °C (Figure 1D).

In conclusion, based on the experimental findings, the optimal conditions for rAAV production are as follows: co-infect Sf9 cells with rBac-SRCR59-Fc/Sn4D-Fc and rBac-RC at a MOI ratio of 0.5:1.0, an initial insect cell density of 2.0 × 10^6 cells/mL, 2% FBS in the insect cell growth medium, and a culture temperature of 30 °C. These optimized conditions will be applied to the subsequent large-scale production of rAAV.

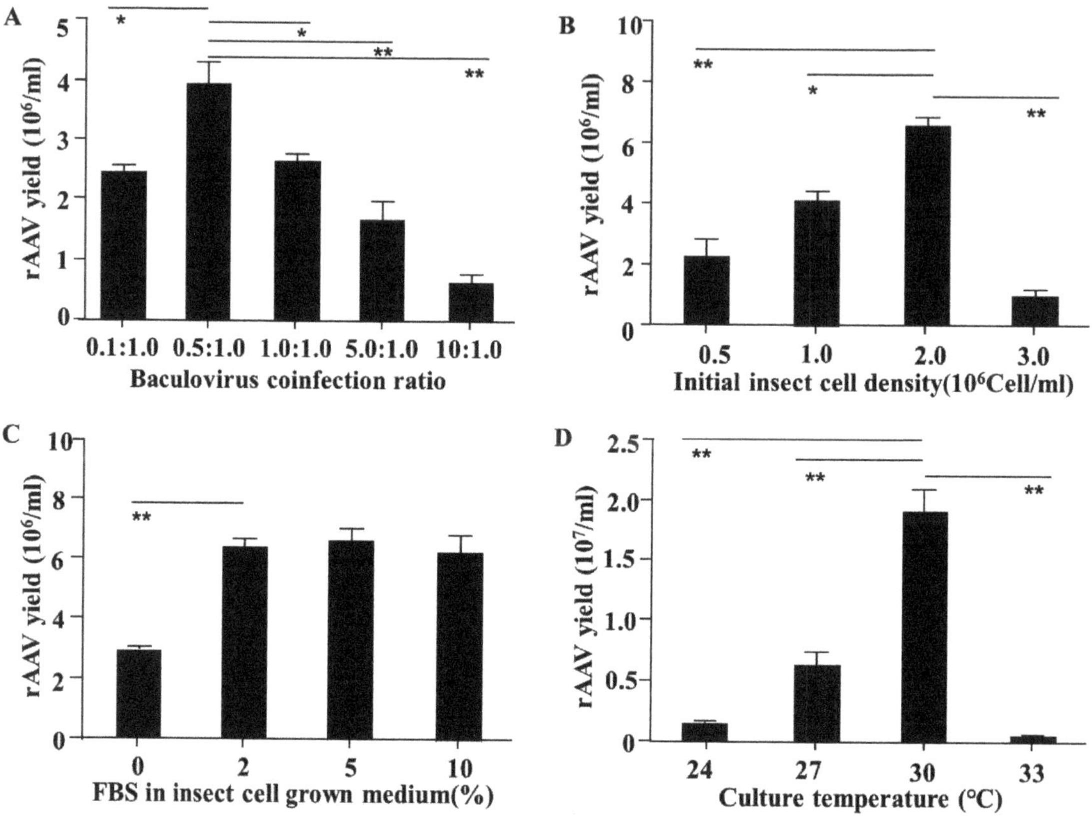

Figure 1. Optimization of rAAV production parameters. (**A**) Effect of baculovirus co-infection ratio (rBac-SRCR59-Fc/Sn4D-Fc:rBac-RC) on rAAV packaging efficiency. Sf9 cells (1.0 × 10^6 cells/mL) were co-infected at MOI ratios of 0.1:1.0, 0.5:1.0, 1.0:1.0, 5.0:1.0, or 10:1.0. The highest rAAV titer (4.0 ± 0.5 × 10^6 IVPs/mL) was achieved at an MOI ratio of 0.5:1.0, which was significantly higher than other ratios ($p < 0.01$ for 5.0:1.0 and 10:1.0; $p < 0.05$ for 1.0:1.0 and 0.1:1.0). (**B**) Impact of initial insect cell density on rAAV yield. Maximum production (6.1 ± 0.1 × 10^6 IVPs/mL) occurred at 2.0 × 10^6 cells/mL, showing significant improvement compared to 0.5 × 10^6 ($p < 0.01$), 3.0 × 10^6 ($p < 0.01$), and 1.0 × 10^6 ($p < 0.05$) cells/mL. (**C**) Influence of fetal bovine serum (FBS) concentration on rAAV production. Optimal yields were observed in 2% FBS medium (6.1 ± 0.1 × 10^6 IVPs/mL), which was significantly higher than serum-free conditions ($p < 0.01$) but showed no significant difference from 5% or 10% FBS groups. (**D**) Effect of culture temperature on rAAV titers. The highest titer (1.9 ± 0.3 × 10^7 IVPs/mL) was achieved at 30 °C, significantly surpassing 27 °C ($p < 0.01$) and 24 °C ($p < 0.01$). No infectious virus was detected at 33 °C. No infectious virus was detected at 33 °C. ** indicates a highly significant difference ($p < 0.01$) and * indicates a significant difference ($p < 0.05$) between experimental groups (two-tailed Student's t-test). Data are presented as mean ± SD from three independent biological replicates. Error bars represent SD.

3.2. Baculovirus-Infected Insect Cells (BIICs) Production

Sf9 cells were cultured in 125 mL shake flasks with 20 mL medium (2 × 10^6 cells/mL) supplemented with 2% FBS. The cells were co-infected with rBac-SRCR59-Fc/Sn4D-Fc (MOI 0.5) and rBac-RC (MOI 1.0) at a temperature of 30 °C. Insect cell density, viability, and rAAV yields were measured at 24, 48, 72, 96, and 120 h post-infection (hpi). As revealed in Figure 2, the highest rAAV titer was observed at 72 hpi (1.9 ± 0.6 × 10^7 IVPs/mL), followed by 48 hpi (1.5 ± 0.5× 10^7 IVPs/mL), 96 hpi (5.2 ± 0.4× 10^6 IVPs/mL), and 120 hpi (0.8 ± 0.6 × 10^6 IVPs/mL). Only small amounts of infectious particles were detected at 24 hpi (2.6 ± 0.5 × 10^2 IVPs/mL). The insect cell viability at 72 hpi, 96 hpi, or 120 hpi was under 90%, while the cell viability remained >90% at 24 or 48 h post-infection (Figure 2). Although the highest rAAV titer was at 72 hpi, considering factors such as maintaining relatively high cell viability and facilitating subsequent operations, 48 hpi was identified as the optimal time for BIICs collection.

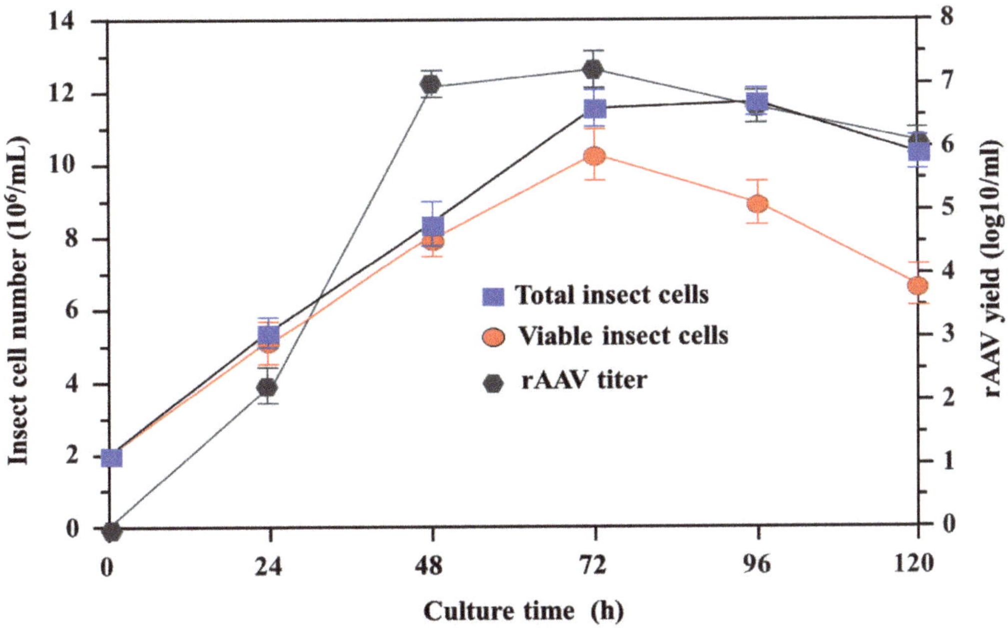

Figure 2. Time course of baculovirus co-infection in the cell flask. To determine the optimum harvest time of BIICs under the optimized baculovirus co-infection conditions, the insect cell density, the number of living cells, and the yields of rAAV were checked at different times after infection.

3.3. High Yields of Recombinant Adeno-Associated Virus Packaging

The large-scale production protocol for rAAV-SRCR59-Fc/Sn4D-Fc is summarized in Figure 3A. The process was scaled up to 2 L insect cell rocking platform bioreactors based on the pilot study of optimization. Sf9 cells (7.5 × 10^5/mL) were cultured in the bioreactor and subjected to fed-batch cultivation at t = −48 h and t = −24 h using fresh 2% FBS insect medium. BIICs were introduced into the final working volume when the cell density reached 2 × 10^6 cells/mL (t = 0). The cells continued to proliferate, and the cell density increased until approximately t = 120 hpi. The highest titer of infectious viral particles (IVPs) of rAAV was achieved at t = 144 hpi (Figure 3B). The rAAV-SRCR59-Fc/Sn4D-Fc was harvested and purified from insect cell bioreactors. The initial yield of rAAV-SRCR59-Fc/Sn4D-Fc was 200 to 600 IVPs/cell and 5.37 ± 0.95 × 10^9 IVPs/mL, with a recovery rate of 80.4 ± 2.9% and a purity of 83.1 ± 5.2%. In contrast, for the production of rAAV in cell flasks, the initial yield was 1–10 IVPs/cell and $3.19 \pm 0.21 \times 10^7$ IVPs/mL, with a recovery rate of $75.3 \pm 3.1\%$ and a purity of $89.7 \pm 1.7\%$, respectively. As shown in Table 2, the comparison between the production in bioreactors and cell flasks clearly demonstrates the superiority of the large-scale production in terms of yield per cell and total IVPs/mL.

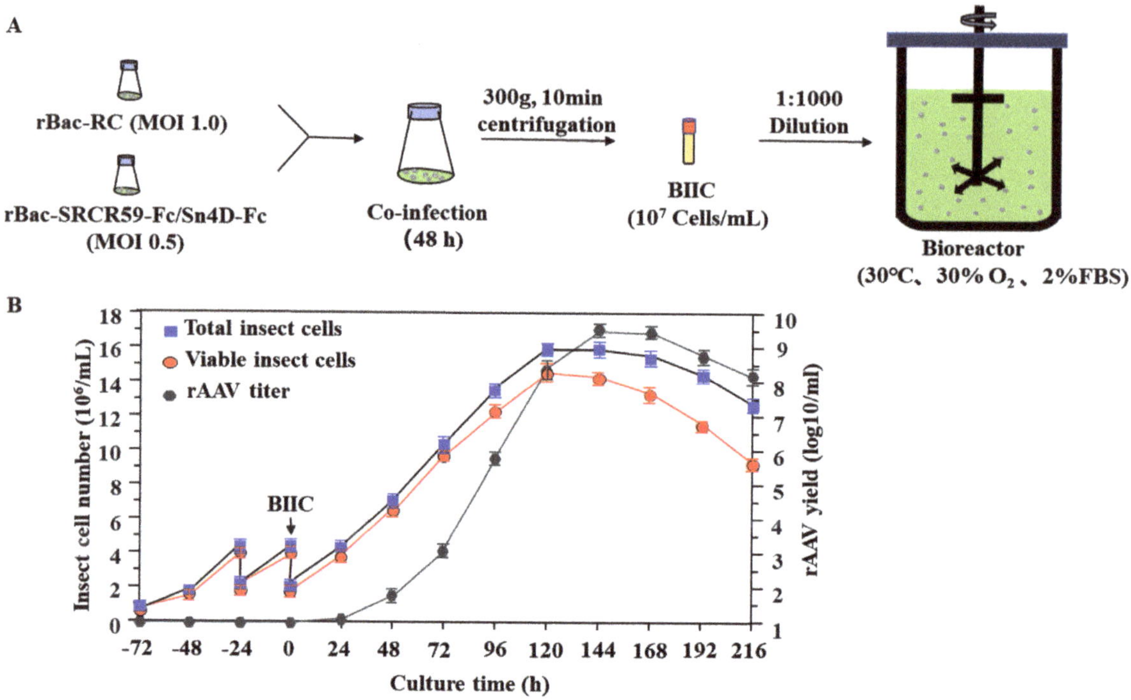

Figure 3. High yields of recombinant adeno-associated virus packaging. (**A**) The protocol of the rAAV-SRCR59-Fc/Sn4D-Fc large-scale production method. (**B**) Time course of rAAV production in the insect cell bioreactor. Sf9 cells (7.5×10^5/mL) were grown in the 2 L bioreactor. Specifically, the cells were fed-batch with fresh medium twice, 48 h and 24 h before infection. BIICs were added to reach the final working volume when the cell density reached 2×10^6 cells/mL (t = 0 hpi). Subsequently, the total insect cell density, viable insect cell density, and rAAV titer were measured every 24 h until 192 h post-infection.

Table 2. A comparison of production and purification performance is conducted between different rAAV packaging strategies.

Packaging Strategy	Yield (IVPs/mL)	Yield (IVPs/Cell)	Recovery (%)	Purity (%)
Cell flask (72 h)	$1.9 \pm 0.3 \times 10^7$	1~10	75.3 ± 3.1	89.7 ± 1.7
Cell bioreactor (144 h)	$5.4 \pm 0.9 \times 10^9$	200~600	80.4 ± 2.9	83.1 ± 5.2

The purification results were determined by analyzing the protein grey value using Image Lab™ software 5.2.1.

3.4. Identification of rAAV Produced and SVR Expressed In Vitro

Electron microscopy demonstrated that the rAAV-SRCR59-Fc/Sn4D-Fc had the typical AAV morphology with a diameter of approximately 22 nm (Figure 4A). Western blotting confirmed that the rAAV reacted positively with an anti-AAV antibody with the standard VP1: VP2: VP3 ratio (Figure 4B). Immunofluorescent analysis demonstrated that 3D4/21 cells infected with rAAV-SRCR59-Fc/Sn4D-Fc were stained with FITC-labeled anti-pig IgG (Figure 4C). In addition to the identification of rAAV, the expression of SVR was also examined. Cell culture media were collected at 48 h, and the two SVR fusions were purified using Protein A Sefinose Columns. Western blotting analysis demonstrated two proteins with the expected molecular weights for SRCR59-Fc (88 kDa) and Sn4D-Fc (72 kDa) (Figure 4D).

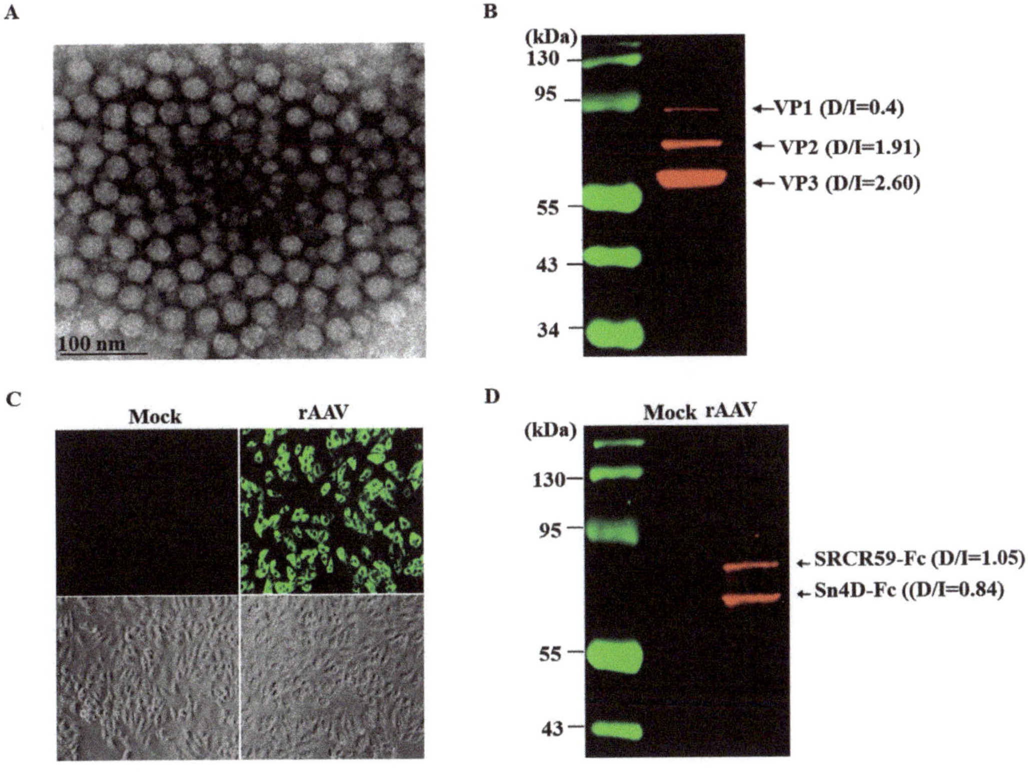

Figure 4. Identification of rAAV particles and SVR expression in rAAV-transduced 3D4/21 cells. (**A**) Purified rAAV-SRCR59-Fc/Sn4D-Fc was stained with 3% phosphotungstic acid and observed under a transmission electron microscope at an acceleration voltage of 75 kV. (**B**) The structural proteins of rAAV were detected by Western blotting using anti-AAV serum as the primary antibody. (**C**) 3D4/21 cells were transduced with rAAV-SRCR59-Fc/Sn4D-Fc and analyzed by immunofluorescence using anti-pig IgG. (**D**) The SVR fusions were purified from the cell medium of rAAV-transduced cells using Protein A affinity columns and analyzed by Western blotting using anti-Sn or anti-CD163 serum. D/I ratios represent target protein densitometry values normalized to loading control using Image J software 1.8.0. Western blot original images can be found in Supplementary Materials.

3.5. Anti-PRRSV Activities of SVR Fusions

The antiviral efficacy of dual soluble viral receptors (SVRs) was evaluated using a Transwell co-culture system (CORNING, Corning, NY, USA) to simulate viral inhibition in primary porcine alveolar macrophages (PAMs). Compared with the empty AAV control group (VR2332: 7.0 ± 0.2 $\log_{10}$ $TCID_{50}$/mL; JXA1: 7.2 ± 0.2 $\log_{10}$ $TCID_{50}$/mL; JS07: 6.9 ± 0.1 $\log_{10}$ $TCID_{50}$/mL; SH1705: 7.6 ± 0.2 $\log_{10}$ $TCID_{50}$/mL), transduction with rAAV-SRCR59-Fc/Sn4D-Fc significantly reduced viral titers across all tested PRRSV-2 strains (Figure 5). Post-treatment titers were VR2332: 2.4 ± 0.2 $\log_{10}$ $TCID_{50}$/mL (4.6 log reduction), JXA1: 2.8 ± 0.4 $\log_{10}$ $TCID_{50}$/mL (4.4 log reduction), JS07: 2.6 ± 0.1 $\log_{10}$ $TCID_{50}$/mL (4.3 log reduction), and SH1705: 3.4 ± 0.3 $\log_{10}$ $TCID_{50}$/mL (4.2 log reduction). On average, viral titers were reduced by 4.3 ± 0.2 $\log_{10}$ $TCID_{50}$/mL ($p < 0.01$, two-tailed Student's t-test). A Kruskal–Wallis test confirmed no significant differences in antiviral responses among

PRRSV-2 strains ($p = 0.12$), indicating broad-spectrum activity of the dual SVRs against genetically diverse isolates.

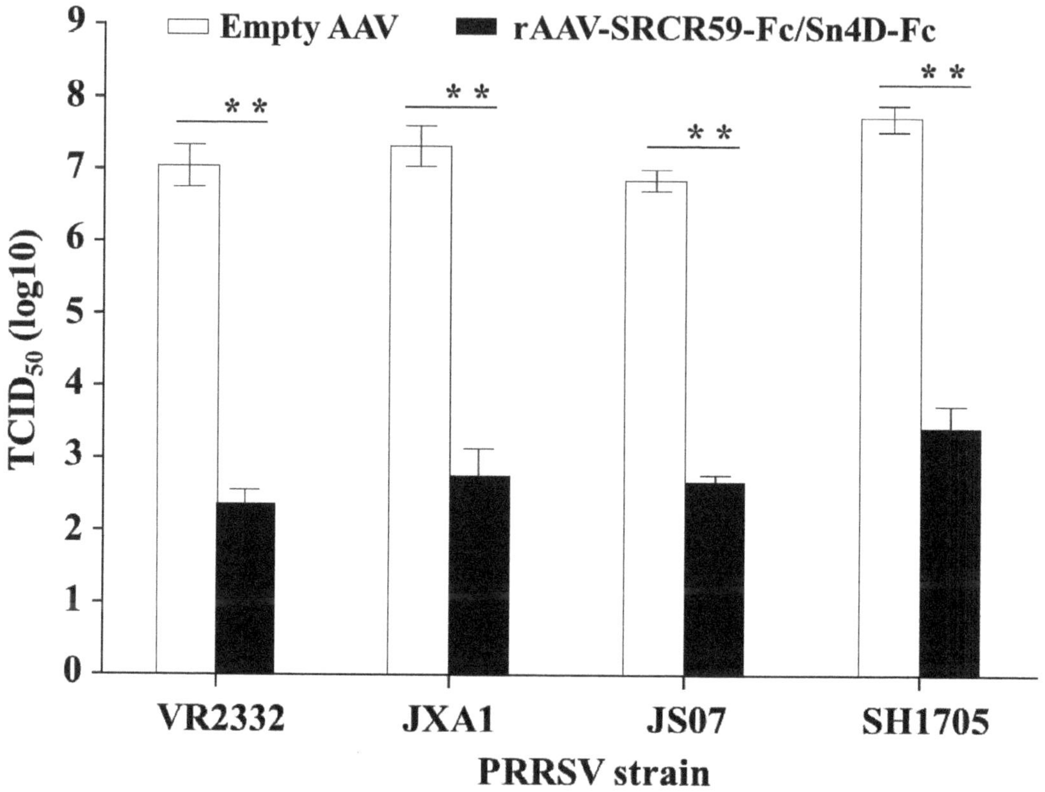

Figure 5. Antiviral activity of rAAV-SRCR59-Fc/Sn4D-Fc against diverse PRRSV-2 strains. 3D4/21 cells were transduced with rAAV-SRCR59-Fc/Sn4D-Fc (MOI 100) or empty AAV control. PAM cells were infected with PRRSV strains (MOI 0.5) and co-cultivated with transduced 3D4/21 cells for 24 h. Viral titers in PAM lysates were determined by $TCID_{50}$ assay on Marc-145 cells. ** indicates highly significant difference ($p < 0.01$) between the empty AAV control and rAAV-SRCR59-Fc/Sn4D-Fc transduction experimental group. Error bars represent SD ($n = 3$).

4. Discussion

The rAAV production system is predicated on the co-infection of two baculoviruses. The co-infection ratio of these two baculoviruses significantly impacts rAAV production. It has been demonstrated that altering the ratios of baculoviruses in the triplicate virtual co-infection system altered the yields of infectious rAAV [30]. Our study found that an MOI ratio of 0.5:1.0 for the two baculoviruses can achieve the highest rAAV yields. This optimal ratio might be associated with a balanced supply of viral components necessary for rAAV assembly. When the ratio exceeds 1.0:1.0, insect cells undergo rapid apoptosis, which undermines rAAV packaging. This could be due to an over-burden of viral genetic material or proteins, triggering cellular stress responses and ultimately leading to cell death. Cell density is a crucial factor influencing rAAV production. Owing to oxygen limitations and challenges in nutrient uptake at high cell densities, optimal cell productivity can be achieved by initiating the infection when the cell density reaches $1–2 \times 10^6$ cells/mL in fresh medium [31]. In our research, we also observed this phenomenon. The highest rAAV yields were obtained from cell flasks with an initial insect cell density of 2.0×10^6 cells/mL, as opposed to 3.0×10^6 cells/mL. At higher cell densities, the competition for oxygen and nutrients intensifies, which may limit the metabolic activity of the cells and consequently reduce rAAV production. Therefore, maintaining an appropriate cell density is essential for efficient rAAV production. The concentration of fetal bovine serum (FBS) in the insect growth medium also plays a role. A high FBS concentration elevates production costs. Moreover, excess nutrients lead to metabolite accumulation that inhibits cell growth. Based on these considerations, the optimal FBS concentration was determined to be 2%. This concentration strikes a balance between providing sufficient nutrients for cell growth and minimizing the negative effects of metabolite accumulation. It also helps in reducing the overall production cost, which is an important aspect for large-scale rAAV production. Temperature is another significant variable. Generally, insect cells are cultured at either 27 °C or 28 °C. Raising the culture temperature to 30 °C enhanced encapsidation, as determined by the ratio of capsids to DNase-resistant particles to infectious particles [32]. In our observations, there was no significant difference in Sf9 cell viability or baculovirus infection efficiency between 27 °C and 30 °C. Nevertheless, rAAV yields at 30 °C were indeed three times higher than those at 27 °C. The increased temperature may accelerate certain enzymatic reactions involved in rAAV assembly, such as the processes of capsid formation and genome encapsidation. However, further studies are needed to fully elucidate the underlying molecular mechanisms. In contrast to prior investigations, our study conducted a comprehensive assessment of multiple factors influencing rAAV production and precisely identified their optimal conditions. These findings not only offer invaluable perspectives for enhancing the efficiency of rAAV production but also establish a solid foundation for the large-scale manufacturing of rAAV.

Studies have indicated that a clinical dose for humans requires 10^{12}–10^{13} rAAV vector particles, depending on the needed therapeutic protein expression level for treatment [33]. Considering the recombinant adeno-associated virus serotype 2 reference standard and the fact that the average human weighs 5–10 times more than a 20-day-old pig, at least 10^9 infectious viral particles per piglet are necessary [34]. In this study, two baculoviruses (rBac-SRCR59-Fc/Sn4D-Fc and rBac-RC) co-infected Sf9 cells in cell flasks under optimized conditions. Unfortunately, the highest yield of rAAV-SRCR59-Fc/Sn4D-Fc was only about 2×10^7 IVPS/mL at 72 h post-infection. This low titer from cell flasks is insufficient for experimental animal models and commercial production. Previous research has demonstrated that the baculovirus–insect cell (BIIC) system and fed-batch strategies are effective for rAAV packaging when three baculoviruses (rBac-GFP, rBac-Rep, or rBac-VP) co-infect insect cells [25,35]. Therefore, we employed a two-baculovirus co-infection system for rAAV-SRCR59-Fc/Sn4D-Fc production. First, we determined the optimal collection time (48 hpi) to optimize the BIIC process. Based on small- and mid-scale pilot studies, we scaled up the process to a 2 L stirred-tank single-use cell bioreactor with temperature and oxygen regulation. We controlled the Sf9 cell density and vitality through a fed-batch approach using a 2% FBS insect growth medium. As a result, the titer yield of rAAV-SRCR59-Fc/Sn4D-Fc in the bioreactor was significantly higher than that in small cell flasks, reaching $5.4 \pm 0.9 \times 10^9$ IVPS/cell, which is sufficient for commercial production and clinical gene therapy in piglets.

Soluble receptors derived from evolutionarily conserved host proteins represent a promising antiviral strategy, particularly for controlling rapidly mutating viruses with diverse serotypes [36,37]. This approach circumvents the need for strain-specific vaccines and addresses the antigenic diversity of PRRSV-2, as demonstrated by our in vitro neutralization assays showing comparable activity of rAAV-delivered Sn4D-Fc and SRCR59-Fc against genetically divergent PRRSV strains (Figure 5). Notably, while murine models validated the sustained SVR expression (35 days) enabled by rAAV vectors, direct translation to porcine hosts warrants caution due to species-specific immune responses [15]. In contrast to rAd vectors, which induced transient SVR persistence (<14 days) in pigs, the enhanced durability of rAAV-SVRs arises from the vector's low immunogenicity and the use of porcine-derived SVR sequences, which minimize neutralizing antibody formation—a critical advantage for repeated veterinary administrations [14]. However, several limitations must be acknowledged. First, in vitro neutralization does not fully recapitulate the complexity of viral pathogenesis in vivo. Second, the long-term safety of rAAV-SVRs in pigs remains uncharacterized, including potential off-target effects and tissue tropism. Future studies should prioritize in vivo challenge experiments, dose optimization, and combination therapies with existing vaccines to maximize protective efficacy. Given the challenges in developing novel PRRS vaccines and the broad-spectrum neutralization capacity of these SVRs, rAAV vectors may offer an efficient alternative for PRRS control.

5. Conclusions

This study establishes an optimized insect cell–baculovirus system for scalable production of rAAV vectors expressing dual soluble viral receptors (SVRs) against PRRSV. Through parameter optimization, high-titer rAAV-SRCR59-Fc/Sn4D-Fc ($5.4 \pm 0.9 \times 10^9$ IVPS/mL) was achieved in a 2 L bioreactor. In vitro assays confirmed efficient SVR expression and broad-spectrum antiviral activity, reducing viral titers of diverse PRRSV strains by ~4.3 log. This platform offers a viable strategy for large-scale rAAV manufacturing, highlighting the potential of rAAV-delivered SVRs as a durable, pan-strain therapeutic to address PRRSV challenges in swine production.

Supplementary Materials: The following supporting information can be downloaded at: https://www.mdpi.com/article/10.3390/vetsci12040366/s1, File S1: Western blot original images.

Author Contributions: Methodology, manuscript writing, X.L. and N.X.; sample collection, software, X.S., L.Z. and Q.S.; laboratory management, conceptualization, data validation, and manuscript polishing, H.S. All authors have read and agreed to the published version of the manuscript.

Funding: This work was supported by the High-level Talents Scientific Research Initiation Project of the Jiangsu Vocational College of Agriculture and Forestry (2024rc21, 2024rc26), the National Key Research and Development Program of China (2018YFC0840400-4), and Priority Academic Program Development of Jiangsu Higher Education Institutions (PAPD).

Institutional Review Board Statement: The experiment was carried out according to the recommendations in the Guide for the Care and Use of Laboratory Animals of Yangzhou University. The protocol was approved by the Medical Experimental Animal Center of Jiangsu Province (Permit Number: NSFC2020-dkxy.02, 27 March 2020).

Informed Consent Statement: Not applicable.

Data Availability Statement: All data generated or analyzed during this study are included in this published article.

Conflicts of Interest: The authors declare no conflicts of interest.

References

1. Neumann, E.J.; Kliebenstein, J.B.; Johnson, C.D.; Mabry, J.W.; Bush, E.J.; Seitzinger, A.H.; Green, A.L.; Zimmerman, J.J. Assessment of the economic impact of porcine reproductive and respiratory syndrome on swine production in the United States. *J. Am. Vet. Med. Assoc.* **2005**, *227*, 385–392. [CrossRef] [PubMed]
2. Terpstra, C.; Wensvoort, G.; Pol, J.M. Experimental reproduction of porcine epidemic abortion and respiratory syndrome (mystery swine disease) by infection with Lelystad virus: Koch's postulates fulfilled. *Vet. Q.* **1991**, *13*, 131–136. [CrossRef]

3. Lunney, J.K.; Benfield, D.A.; Rowland, R.R. Porcine reproductive and respiratory syndrome virus: An update on an emerging and re-emerging viral disease of swine. *Virus Res.* **2010**, *154*, 1–6.
4. Pejsak, Z.; Stadejek, T.; Markowska-Daniel, I. Clinical signs and economic losses caused by porcine reproductive and respiratory syndrome virus in a large breeding farm. *Vet. Microbiol.* **1997**, *55*, 317–322. [CrossRef]
5. Butler, J.E.; Lager, K.M.; Golde, W.; Faaberg, K.S.; Sinkora, M.; Loving, C.; Zhang, Y.I. Porcine reproductive and respiratory syndrome (PRRS): An immune dysregulatory pandemic. *Immunol. Res.* **2014**, *59*, 81–108. [PubMed]
6. Charemtantanakul, W. Porcine reproductive and respiratory syndrome virus vaccines: Immunogenicity, efficacy and safety aspects. *World J. Virol.* **2012**, *1*, 23–30. [CrossRef]
7. Kimman, T.G.; Cornelissen, L.A.; Moormann, R.J.; Rebel, J.M.; Stockhofe-Zurwieden, N. Challenges for porcine reproductive and respiratory syndrome virus (PRRSV) vaccinology. *Vaccine* **2009**, *27*, 3704–3718. [PubMed]
8. Kappes, M.A.; Faaberg, K.S. PRRSV structure, replication and recombination: Origin of phenotype and genotype diversity. *Virology* **2015**, *479–480*, 475–486.
9. Shi, M.; Lam, T.T.; Hon, C.C.; Hui, R.K.; Faaberg, K.S.; Wennblom, T.; Murtaugh, M.P.; Stadejek, T.; Leung, F.C. Molecular epidemiology of PRRSV: A phylogenetic perspective. *Virus Res.* **2010**, *154*, 7–17.
10. Van Breedam, W.; Delputte, P.L.; Van Gorp, H.; Misinzo, G.; Vanderheijden, N.; Duan, X.; Nauwynck, H.J. Porcine reproductive and respiratory syndrome virus entry into the porcine macrophage. *J. Gen. Virol.* **2010**, *91*, 1659–1667. [CrossRef]
11. Van Gorp, H.; Van Breedam, W.; Van Doorsselaere, J.; Delputte, P.L.; Nauwynck, H.J. Identification of the CD163 protein domains involved in infection of the porcine reproductive and respiratory syndrome virus. *J. Virol.* **2010**, *84*, 3101–3105. [CrossRef] [PubMed]
12. Van Breedam, W.; Van Gorp, H.; Zhang, J.Q.; Crocker, P.R.; Delputte, P.L.; Nauwynck, H.J. The M/GP5 glycoprotein complex of porcine reproductive and respiratory syndrome virus binds the sialoadhesin receptor in a sialic acid-dependent manner. *PLoS Pathog.* **2010**, *6*, e1000730. [CrossRef]
13. Chen, Y.; Guo, R.; He, S.; Zhang, X.; Xia, X.; Sun, H. Additive inhibition of porcine reproductive and respiratory syndrome virus infection with the soluble sialoadhesin and CD163 receptors. *Virus Res.* **2014**, *179*, 85–92. [CrossRef]
14. Xia, W.; Wu, Z.; Guo, C.; Zhu, S.; Zhang, X.; Xia, X.; Sun, H. Recombinant adenovirus-delivered soluble CD163 and sialoadhesin receptors protected pigs from porcine reproductive and respiratory syndrome virus infection. *Vet. Microbiol.* **2018**, *219*, 1–7. [CrossRef]
15. Liu, X.; Xia, W.; Zhang, X.; Xia, X.; Sun, H. Fusion expression of the two soluble viral receptors of porcine reproductive and respiratory syndrome virus with a single adeno-associated virus vector. *Res. Vet. Sci.* **2021**, *135*, 78–84. [CrossRef] [PubMed]
16. Daya, S.; Berns, K.I. Gene therapy using adeno-associated virus vectors. *Clin. Microbiol. Rev.* **2008**, *21*, 583–593. [CrossRef] [PubMed]
17. Mitchell, A.M.; Nicolson, S.C.; Warischalk, J.K.; Samulski, R.J. AAV's anatomy: Roadmap for optimizing vectors for translational success. *Curr. Gene Ther.* **2010**, *10*, 319–340. [CrossRef]
18. Bish, L.T.; Morine, K.; Sleeper, M.M.; Sanmiguel, J.; Wu, D.; Gao, G.; Wilson, J.M.; Sweeney, H.L. Adeno-associated virus (AAV) serotype 9 provides global cardiac gene transfer superior to AAV1, AAV6, AAV7, and AAV8 in the mouse and rat. *Hum. Gene Ther.* **2008**, *19*, 1359–1368. [CrossRef]
19. Qi, Y.F.; Li, Q.H.; Shenoy, V.; Zingler, M.; Jun, J.Y.; Verma, A.; Katovich, M.J.; Raizada, M.K. Comparison of the transduction efficiency of tyrosine-mutant adeno-associated virus serotype vectors in kidney. *Clin. Exp. Pharmacol. Physiol.* **2013**, *40*, 53–55. [CrossRef]
20. Urabe, M.; Ding, C.; Kotin, R.M. Insect cells as a factory to produce adeno-associated virus type 2 vectors. *Hum. Gene Ther.* **2002**, *13*, 1935–1943. [CrossRef]
21. Mietzsch, M.; Casteleyn, V.; Weger, S.; Zolotukhin, S.; Heilbronn, R. OneBac 2.0: Sf9 Cell Lines for Production of AAV5 Vectors with Enhanced Infectivity and Minimal Encapsidation of Foreign DNA. *Hum. Gene Ther.* **2015**, *26*, 688–697. [CrossRef] [PubMed]
22. Wensvoort, G.; Terpstra, C.; Pol, J.M.; ter Laak, E.A.; Bloemraad, M.; de Kluyver, E.P.; Kragten, C.; van Buiten, L.; den Besten, A.; Wagenaar, F. Mystery swine disease in The Netherlands: The isolation of Lelystad virus. *Vet. Q.* **1991**, *13*, 121–130. [CrossRef] [PubMed]
23. Tian, K.; Yu, X.; Zhao, T.; Feng, Y.; Cao, Z.; Wang, C.; Hu, Y.; Chen, X.; Hu, D.; Tian, X.; et al. Emergence of fatal PRRSV variants: Unparalleled outbreaks of atypical PRRS in China and molecular dissection of the unique hallmark. *PLoS ONE* **2007**, *2*, e526. [CrossRef]
24. Lin, H.; Zhe, M.; Xin, H.; Lei, C.; Fan, H. Construction and immunogenicity of a recombinant swinepox virus expressing a multi-epitope peptide for porcine reproductive and respiratory syndrome virus. *Sci. Rep.* **2017**, *7*, 43990.
25. Cecchini, S.; Virag, T.; Kotin, R.M. Reproducible high yields of recombinant adeno-associated virus produced using invertebrate cells in 0.02-to 200-liter cultures. *Hum. Gene Ther.* **2011**, *22*, 1021–1030. [CrossRef] [PubMed]
26. Kuwae, S.; Ohda, T.; Tamashima, H.; Miki, H.; Kobayashi, K. Development of a fed-batch culture process for enhanced production of recombinant human antithrombin by Chinese hamster ovary cells. *J. Biosci. Bioeng.* **2005**, *100*, 502–510. [CrossRef]

27. Matsushita, T.; Elliger, S.; Elliger, C.; Podsakoff, G.; Villarreal, L.; Kurtzman, G.J.; Iwaki, Y.; Colosi, P. Adeno-associated virus vectors can be efficiently produced without helper virus. *Gene Ther.* **1998**, *5*, 938–945. [CrossRef]
28. Lee, Y.J.; Park, C.K.; Nam, E.; Kim, S.H.; Lee, O.S.; Lee, D.S.; Lee, C. Generation of a porcine alveolar macrophage cell line for the growth of porcine reproductive and respiratory syndrome virus. *J. Virol. Methods* **2010**, *163*, 410–415. [CrossRef]
29. Jacobs, A.C.; Hermann, J.R.; Munoz-Zanzi, C.; Prickett, J.R.; Roof, M.B.; Yoon, K.J. Stability of porcine reproductive and respiratory syndrome virus at ambient temperatures. *J. Vet. Diagn. Investig* **2010**, *22*, 257–260. [CrossRef]
30. Meghrous, J.; Aucoin, M.G.; Jacob, D.; Chahal, P.S.; Arcand, N.; Kamen, A.A. Production of recombinant adeno–associated viral vectors using a baculovirus/insect cell suspension culture system: From shake flasks to a 20-L bioreactor. *Biotechnol. Prog.* **2005**, *21*, 154–160. [CrossRef]
31. Carinhas, N.; Bernal, V.; Yokomizo, A.Y.; Carrondo, M.J.; Oliveira, R.; Alves, P.M. Baculovirus production for gene therapy: The role of cell density, multiplicity of infection and medium exchange. *Appl. Microbiol. Biotechnol.* **2009**, *81*, 1041–1049. [CrossRef] [PubMed]
32. Aucoin, M.G.; Perrier, M.; Kamen, A.A. Improving AAV vector yields in insect cells by modulating the temperature after infection. *Biotechnol. Bioeng.* **2007**, *97*, 1501–1509. [CrossRef] [PubMed]
33. Grimm, D.; Kleinschmidt, J.A. Progress in adeno- associated virus type 2 vector production: Promises and prospects for clinical use. *Hum. Gene Ther.* **1999**, *10*, 2445–2450. [CrossRef] [PubMed]
34. Lock, M.; McGorray, S.; Auricchio, A.; Ayuso, E.; Beecham, E.J.; Blouin-Tavel, V.; Bosch, F.; Bose, M.; Byrne, B.J.; Caton, T.; et al. Characterization of a recombinant adeno-associated virus type 2 Reference Standard Material. *Hum. Gene Ther.* **2010**, *21*, 1273–1285. [CrossRef]
35. Liu, Y.K.; Yang, C.J.; Liu, C.L.; Shen, C.R.; Shiau, L.D. Using a fed-batch culture strategy to enhance rAAV production in the baculovirus/insect cell system. *J. Biosci. Bioeng.* **2010**, *110*, 187–193. [CrossRef]
36. Christiansen, G.J.; Nielsen, J.; Madsen, J.S. Soluble receptors as antiviral agents. *Curr. Opin. Biotechnol.* **2000**, *11*, 448–453.
37. Fechner, H.; Pinkert, S.; Geisler, A.; Poller, W.; Kurreck, J. Pharmacological and biological antiviral therapeutics for cardiac coxsackievirus infections. *Molecules* **2011**, *16*, 8475–8503. [CrossRef]

veterinary sciences

MDPI

Article

The Detection of Mixed Infection with Canine Parvovirus, Canine Distemper Virus, and Rotavirus in Giant Pandas by Multiplex PCR

Ai Liu [1,2,†], Wenyue Qiao [1,2,†], Rui Ma [3], Qigui Yan [1,2], Shan Zhao [1,2] and Yifei Lang [1,2,*]

[1] College of Veterinary Medicine, Sichuan Agricultural University, Chengdu 611130, China
[2] Chengdu National Agricultural Science and Technology Center, Chengdu 610213, China
[3] Sichuan Key Laboratory of Conservation Biology for Endangered Wildlife, Chengdu Research Base of Giant Panda Breeding, 1375 Panda Road, Chenghua District, Chengdu 610081, China
* Correspondence: y_langviro@163.com
† These authors contributed equally to this work.

Simple Summary: Virus infections pose significant threats to the health and survival of giant pandas, an endangered species, making rapid and accurate diagnosis crucial for effective management and conservation efforts. In this study, we developed a multiplex PCR (mPCR) method to detect canine parvovirus type 2 (CPV-2), canine distemper virus (CDV), and giant panda rotavirus (GPRV) in pandas. The mPCR approach was validated for its high sensitivity and specificity in identifying these viruses, including mixed infections. In a sample of 218 giant pandas, CPV-2, CDV, and GPRV were detected in 19.72%, 7.34%, and 6.42% of the cases, respectively, with over half of the positive samples showing mixed infections. The findings were further confirmed using sequencing and phylogenetic analysis. This method offers a valuable tool for clinical diagnosis and epidemiological studies, aiding in the protection of the giant panda population.

Abstract: The well-being and subsistence of giant pandas, an endangered species with a limited distribution, are currently threatened by a number of viruses, including canine parvovirus (CPV-2), canine distemper virus (CDV), and giant panda rotavirus (GPRV). To allow for timely intervention upon viral infection, it is necessary to execute rapid and accurate diagnosis of potential mixed viral infections. In the present study, we developed and validated a multiplex PCR (mPCR) approach for the detection of CPV-2, CDV, and GPRV infections. The results indicate that the method could selectively amplify the three viruses with high sensitivity and specificity, which are necessary attributes in clinical settings. Utilizing the established method, (sub)clinical giant panda samples were examined, and CPV-2, CDV, and GPRV were found in 19.72% (43 out of 218), 7.34% (16 out of 218), and 6.42% (14 out of 218) of the samples, respectively. Noticeably, mixed infections of two or three viruses were common, and this was generally observed in CDV- or GPRV-positive samples. Meanwhile, mPCR results were further validated with sequencing and the phylogenetic analysis of full-length sequences of viral genes. Taken together, our study provides an approachable assay which enables the quick detection of the three viruses mentioned above, which will benefit clinical diagnosis and laboratory epidemiological-based investigations of the giant panda population.

Keywords: canine parvovirus (CPV-2); canine distemper virus (CDV); giant panda rotavirus (GPRV); multiplex PCR; giant panda

 https://doi.org/10.3390/vetsci12020081

1. Introduction

Viruses are obligate intracellular parasites that can cause infectious diseases in different plant and animal species [1]. Unable to reproduce by themselves, their propagation is fully reliant on the metabolic and biosynthetic processes in living host cells [2,3]. To this end, viral infection often conveys detrimental consequences to their host and was even responsible for several major pandemics in human history. Meanwhile, viruses do not infect just one particular host; on the contrary, they tend to infect many host species within reach [4]. The host range expansion of viruses, mediated by cross-species transmission, provides a certain basis for the emergence of new diseases [5–7]. Such characteristics raise public health concerns and also pose an inevitable threat to a variety of endangered wild animals [8], such as giant pandas (*Ailuropoda melanoleuca*) [9,10].

As a vulnerable species listed by the International Union for the Conservation of Nature (IUCN), giant pandas appear as touristic icons and key representatives of biodiversity conservation. Today, giant pandas are facing many challenges, including habitat fragmentation, reproduction inconveniency, and numerous bacteria and virus infections [9–13]. Over the years, many viruses were confirmed to be able to infect giant panda, including canine parvovirus type 2 (CPV-2), canine adenovirus type 1 (CAV-1), canine distemper virus (CDV), giant panda rotavirus (GPRV), and influenza A virus (IAV). These lead to various and even fatal consequences [14–18]. Recently, with the advances in viromics research, traces of infection with more virus species were identified in giant pandas, as detailed in Figure 1 (Supplementary Table S1). Therefore, the etiological and epidemiological surveillance of viral infections that endanger the situation of giant pandas should be reinforced regularly to provide early warning for forthcoming threats.

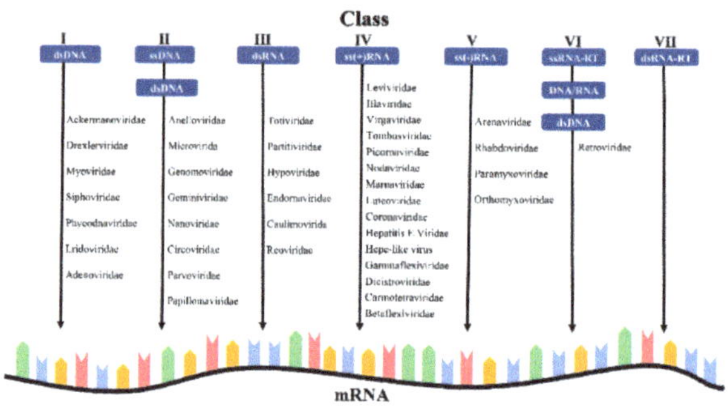

Figure 1. The Baltimore viral categorization system is used to show the various classes of viruses which infect giant pandas. I: Classes of double-stranded DNA viruses (dsDNA); II: classes of single-stranded DNA viruses (ssDNA); III: classes of double-stranded RNA viruses (dsRNA); IV: classes of positive-sense single-stranded RNA viruses (+ssRNA); V: classes of negative-sense single-stranded RNA viruses (-ssRNA); VI: classes of single-stranded RNA reverse transcription viruses (ssRNA-RT); VII: classes of double-stranded RNA reverse transcription viruses (dsRNA-RT).

In the present study, we confirmed the widespread infection and sporadic co-infection of CPV-2, CDV, and GPRV in giant pandas via the establishment and validation of a multiplex polymerase chain reaction (mPCR) assay. Our findings highlight the potential threat of these viruses to giant pandas, emphasize the importance of the timely and systematic surveillance of these viruses, and provide a fast and efficient method to achieve this goal.

2. Materials and Methods

2.1. Viruses and Cells

The CDV (strain A75/17, accession number: AF164967.1), CPV-2 (strain CPV-LZ2, accession number: JQ268284.1) and GPRV (strain CH-1, accession number: GU205762.1) samples used in this study were retrieved from the virus database in our laboratory. CDV and GPRV were propagated in porcine kidney epithelial cells (LLC-PK1), while CPV-2 was propagated in Crandell–Rees Feline Kidney (CRFK) cells. All cells were maintained in Dulbecco's modified Eagle medium (DMEM, Gibco) supplemented with 10% fetal bovine serum (FCS), penicillin (100 IU/mL) and streptomycin (100 µg/mL). Porcine kidney epithelial (LLC-PK1) cells and Crandell–Rees Feline Kidney (CRFK) cells were purchased from ATCC (American Type Culture Collection, Rockefeller, Maryland, MD, USA).

2.2. Fecal Samples from Giant Panda

Giant panda fecal samples (n = 218) were collected from giant pandas at Chengdu Research Base of Giant Panda Breeding, China, between the year 2015 and 2020. Most of the sampled pandas are healthy, with the exception of approximately 10% of them, which displayed signs such as mild diarrhea, anorexia, and vomiting.

2.3. Primer Design

Primers for the multiplex PCR assay were designed based on the CDV N gene, the GPRV NSP1 gene, and the CPV-2 NS1 gene. Representative viral sequences of each virus were first analyzed with multiple-sequence alignment using MEGA X, and primers were selected from the conservative regions within each gene. The selected primer candidates were then analyzed with a BLAST search (NCBI) and the DNASTAR Lasergene software 15.1 package to omit homology with the giant panda genome or other viral sequences. To allow for further analysis of positive samples, sets of primers were also designed to amplify the corresponding full-length coding sequences of the CDV N gene, the GPRV NSP1 gene, and the CPV-2 NS1 gene. All the primers were synthesized in Sangon Biotech Company (Shanghai, China), and full sequence information is detailed in Tables 1 and 2.

Table 1. Primers used for the mPCR method.

Virus	Primer Name	Target	Sequence (5′-3′)	PCR Products (bp)
CDV	CDV-N-F	N	GCATGTCATTATAGTCCTAATCC	757
	CDV-N-R		AAGAGCCGGATACATAGTT	
GPRV	GPRV-NSP1-F	NSP1	AATGATACATGGAGACCATCA	432
	GPRV-NSP1-R		AGTCTGATTAACATTGTTCATGT	
CPV-2	CPV-NS-F	NS1	ACAACGTCCAACTAAATGGA	293
	CPV-NS-R		ACAATGCCAGCCTTGAT	

Table 2. Primers used for the full-length gene sequences.

Virus	Primer Name	Target	Sequence (5′-3′)	PCR Products (bp)
CDV	CDV-H-F1	H	TTAGGGCTCAGGTAGTCCA	915
	CDV-H-R1		TCTACACACAAGGAAGCCA	
	CDV-H-F2		ACTCCAGACAACCAACTAT	1063
	CDV-H-R2		CTAAGTCCAATTGAAATGTGT	
GPRV	GPRV-VP3-F1	VP3	GGCTATTAAAGCAGTACTAG	1581
	GPRV-VP3-R1		AAGAATGTCCAGTCATAGA	
	GPRV-VP3-F2		CCAATGATTACATAGTAGC	1177
	GPRV-VP3-R2		CATCATGACTAGTGTGTTAAG	
CPV-2	CPV-VP2-F	VP2	CCACCACCTCATATTTTCAT	1936
	CPV-VP2-R		CCTTCTAAATCCTATATCAAGTACAA	

2.4. Preparation of Standard Plasmids

DNA sequences representing the ideal mPCR products of CDV (757 bp), GPRV (432 bp), and CPV-2 (293 bp) were synthesized and ligated into the pMD19-T Vector (TaKaRa). All standard plasmids were validated via conventional bidirectional sanger sequencing.

2.5. Establishment of the mPCR Method

The linearized standard plasmids (pMD-CDV-N, pMD-GPRV-NSP1, and pMD-CPV-2-NS) were utilized as architype templates to allow for the optimization of the PCR annealing temperature (Ta). The concentration of each primer was fixed at 100 nM, while each primer constituted 0.25 μL of the reaction volume (primer information is detailed in Table 1). The multiplex PCR assay was then conducted in a total volume of 20 μL per reaction, containing 10 μL of 2 × Taq Master Mix (Vazyme, P112-01), mixed template (pMD-CDV-N, pMD-GPRV-NSP1 and pMD-CPV-2-NS, 10 ng in total), primer mix, and ddH2O. The annealing temperatures for the mPCR reaction were optimized by ramping the temperature from 45 °C to 55 °C, while increasing by 1 °C in a sequential order. The procedure was as follows: 95 °C for 3 min, followed by 34 cycles of 95 °C for 15 s, the diversion Ta for 15 s, and cycles at 72 °C for 15 s, with a final extension at 72 °C for 5 min. After completion, 5 μL of the mPCR product was evaluated by 1.0% agarose gel electrophoresis.

2.6. Specificity and Sensitivity Analysis of the Optimized mPCR Assay

The optimized mPCR assay was used to examine the specificity of the multiplex PCR amplification of all possible combinations, namely, CDV, GPRV, CPV-2, CDV + GPRV, CDV + CPV-2, GPRV + CPV-2, and CDV + GPRV + CPV-2. The amplified target bands were then recovered, purified, and sequenced after ligation to pMD19-T Vector (TaKaRa) to confirm the specificity of the multiplex PCR. To further test the mPCR sensitivity, the concentrations of standard plasmids (pMD-CDV-N, pMD-GPRV-NSP1 and pMD-CPV-2-NS) were adjusted to 10 ng/μL as measured by a NanoDrop 2000 UV-vis spectrophotometer (Thermo Scientific, Waltham, MA, USA). The copy number was then calculated using the following formula: copies/μL= NA (copies/mol) × concentration (g/μL)/MW (g/mol). Moreover, each standard plasmid was serially diluted 10-fold, starting from 1×10^8 to 1×10^1 copies/μL with ddH2O. The plasmids of each dilution were then used as templates for the sensitivity testing of multiplex PCR.

2.7. Viral Nucleic Acid Extraction and Reverse Transcription

Total nucleic acid was extracted from giant panda fecal samples using a DNA/RNA extraction kit (Vazyme, RM201-02) and Stool DNA Isolation Kit (Foregene, DE-05713). This was performed according to the manufacturer's protocol. The extracted total nucleic acids were divided into two portions; one was used for the amplification of the DNA viruses (CPV-2), and the other was used for the reverse transcription (RT) and amplification of the RNA viruses (CDV and GPRV). The RNA extracted from samples was reverse-transcribed to cDNA using the 2×RT OR-EasyTM Mix (Foregene, RT-01022) according to the manufacturer's instructions. The cDNA and DNA were stored at −20 °C until PCR amplification, while the isolated RNA was stored at −80 °C.

2.8. Evaluation of Clinical Samples

A total of 218 panda fecal samples were collected and stored at −80 °C until use. The established multiplex PCR method was then used for virus detection using the isolated DNA and cDNA samples, which were obtained as described above. After mPCR amplification and 1.0% agarose gel electrophoresis, the infection rate was calculated and evaluated.

2.9. Extensive Full-Length Viral Gene Sequencing and Phylogenetic Analysis

To provide further detailed information about the positive samples, the DNA of the DNA virus (CPV-2) and the cDNA of the RNA viruses (CDV, GPRV) of mPCR-positive samples were extracted for PCR amplification. Sequence information about viral structural proteins, namely CDV-H, GPRV-VP3, and CPV-2-VP2, was obtained with the primer pairs listed in Table 2. The reaction was performed in a 20 μL volume that contained 10 μL of 2 × Taq Master Mix (Vazyme, P112-01), 1 μL each of the different templates described above, 2 μL primer sets (final concentration of each primer is 100μM), and 7 μL of ddH2O. The procedure was as follows: 95 °C for 3 min, followed by 34 cycles of 95 °C for 15 s, 47 °C for 15 s, and 72 °C for 15 s, with a final extension at 72 °C for 5 min. The PCR fragments obtained were cloned in pMD19-T Vector (TaKaRa), and different clones were analyzed by conventional bidirectional sanger sequencing. The sequencing results were then assembled by the DNASTAR Lasergene software 15.1 package and processed by the online tool Clustal Omega (https://www.ebi.ac.uk/Tools/msa/clustalo/, accessed on 15 January 2023). Multiple-alignment and phylogenetic analyses of the full-length CDV-H, GPRV-VP3, and CPV-2-VP2 genes were then conducted with representative viral sequences from Genbank using the neighbor-joining method in MEGA X, respectively.

3. Results

3.1. Establishment of a Multiplex PCR (mPCR) Method That Allows Detection of CDV, CPV-2 and GPRV

Through the cautious design of primers and the adjustment of PCR reaction conditions, the specific detection of the three viruses listed above could be achieved. Plasmid-based viral templates, alone or in combination, were first used to verify the mPCR setting, while the products were separated on a 1.0% agarose gel (Figure 2). The results indicated that specific PCR bands (757 bp for CDV, 432 bp for GPRV, and 293 bp for CPV-2) were clearly visible in each group without unspecific amplification. The authenticity of each PCR product was further confirmed by T-A cloning and sequencing analysis. Meanwhile, to test mPCR's compatibility with different DNA polymerases, we tested the mPCR system at different annealing temperatures. As shown in Figure 3, clear PCR bands were equally visible with annealing temperatures ranging from 43 °C to 52 °C, and a temperature of 48 °C was used for further analysis.

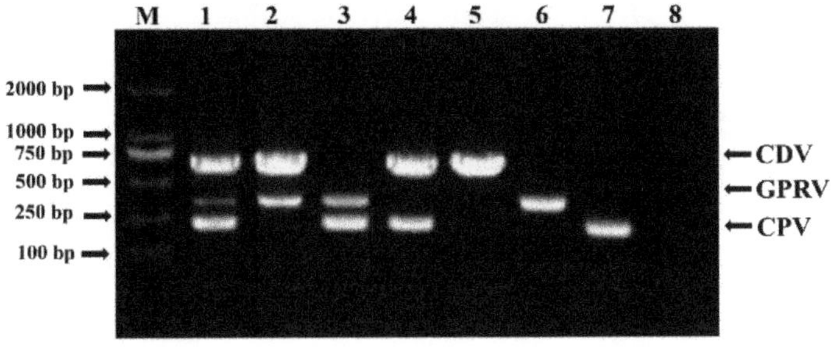

Figure 2. Specificity detection using the mPCR method. M: DL 2000 DNA Marker; 1: CDV + GPRV +CPV-2; 2: CDV + GPRV; 3: GPRV + CPV-2; 4: CDV +CPV-2; 5: CDV; 6: GPRV; 7: CPV-2; 8: negative control.

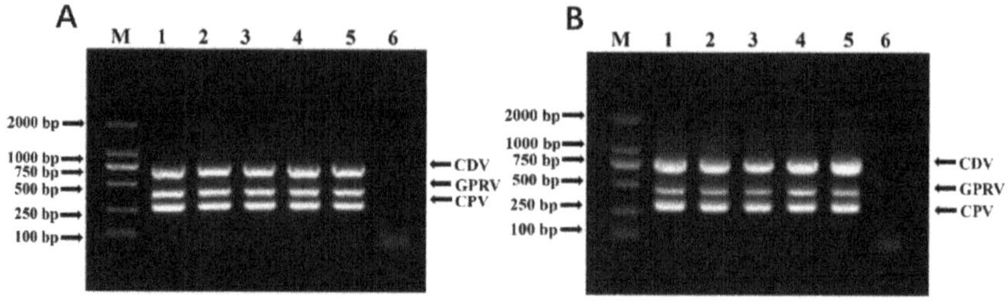

Figure 3. Optimal annealing temperatures for mPCR method. (**A**) M: DL 2000 bp DNA Marker; 1–5: gradient annealing temperatures were 43 °C, 44 °C, 45 °C, 46 °C, and 47 °C, respectively; 6: negative control. (**B**) M: DL 2000 bp DNA Marker; 1–5: gradient annealing temperatures were 48 °C, 49 °C, 50 °C, 51 °C, and 52 °C, respectively; 6: negative control.

3.2. Sensitivity of the Established mPCR Method

To explore the detection limit of the established mPCR method, different concentrations of plasmids, alone or in combination, were employed as templates for the mPCR reaction. The results showed that the minimum detection limit for the mPCR was 1×10^3 viral copies of each virus when the mixed plasmids were used as the template (Figure 4); however, when only one plasmid was used as template, the sensitivity of the method was higher with minimum detection limits for CDV of 1×10^1 viral copies, and for GPRV and CPV-2 of 1×10^2 viral copies, respectively (Figure 4).

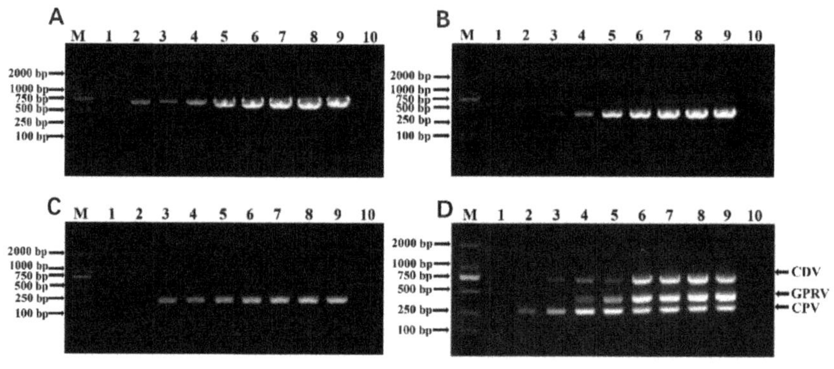

Figure 4. The sensitivity of the established mPCR method. The three single plasmids (pMD-CDV-N, pMD-GPRV-NSP1 and pMD-CPV-2-NS), diluted from 1×10^8 to 1×10^1 copies/μL, and the mixed plasmids (pMD-CDV-N/pMD-GPRV-NSP1/pMD-CPV-2-NS), diluted from 1×10^8 to 1×10^1 copies/μL, were used to determine the minimum detection limit of the mPCR method. (**A**) The sensitivity of pMD-CDV-N; (**B**) the sensitivity of pMD-GPRV-NSP1; (**C**) the sensitivity of pMD-CPV-2-NS; (**D**) the sensitivity of pMD-CDV-N/pMD-GPRV-NSP1/pMD-CPV-2-NS. M: DL 2000 DNA Marker; 1–9: 1×10^1–1×10^8 copies/μL; 10: negative control.

3.3. Detection of Viral Co-Infections in Giant Panda from Fecal Samples

Using the mPCR method established, a total of 218 hiant panda fecal samples were analyzed to detect viral infection. As indicated in Figure S1, the PCR electrophoresis pattern of the clinical samples was similar to that obtained with plasmid templates, where clear-cut bands were visible for each virus and each possible combination, and occurrences of bands with other sizes were unseen throughout the detection of all samples. The analysis of (sub)clinical samples indicated that 7.34% (16 out of 218), 6.42% (14 out of 218), and 19.72% (43 out of 218) of the samples were positive for CDV, GPRV, and CPV-2, respectively (Figure 5A). Venn analysis revealed the true complex nature of viral co-infections. As shown in Figure 5B, over half of CDV (n = 8)- or GPRV (n = 9)-positive samples were also positive for CPV-2, while one CP-2V-negative sample was positive for both CDV and GPRV. Noticeably, 1.38% of the total samples (3 out of 218) were positive for all three viruses, indicating that superinfections might have occurred in those pandas at the time of sampling.

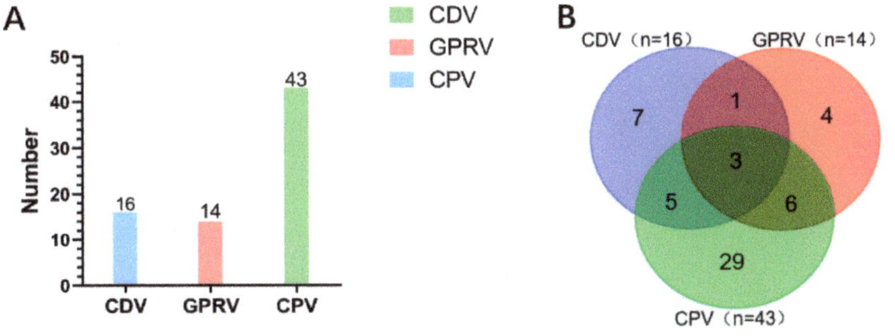

Figure 5. The evaluation of (sub)clinical samples. The detection of viral co-infection in giant panda fecal sample by mPCR (n = 218). (**A**) The number of positive samples is graphed for the mPCR method performed. (**B**) The overlap of viral infections is shown in a Venn diagram.

3.4. Consolidation and Validation of mPCR Results Through Regular PCR and Phylogenetic Analysis

To further verify the specificity of the mPCR results, we designed primers to amplify the full-length sequence of the CDV H gene, the GPRV VP3 gene, and the CPV-2 VP2 gene, while full sequence information is detailed in Table 2. Regular PCR analyses were performed on randomly selected positive samples (n = 10), with the exception of the samples (n = 3) that were positive for all three viruses in the mPCR analysis. The results of the mPCR analysis were well correlated with routine PCR regarding sample positivity, while the sequencing analysis of routine PCR products converged to one or two represented sequences per virus, namely, n = 1 for CDV, n = 1 for GPRV, and n = 2 for CPV-2. Next, phylogenetic analysis was performed with the complete gene sequences obtained. Figure 6 shows the viral sequence clades with major field isolates in dogs (CDV and CPV-2) or pigs (GPRV) that are prevalent in Chinese companion animals or the breeding industry.

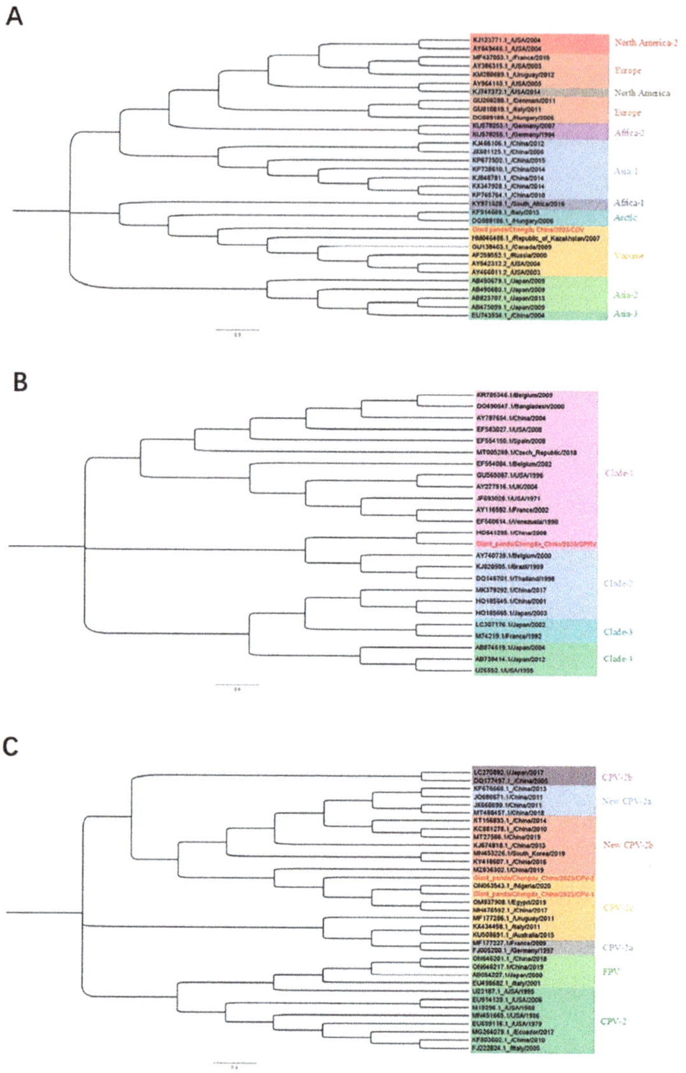

Figure 6. (**A**) Phylogenetic analysis of canine distemper virus (CDV) strains based on H gene sequences; (**B**) phylogenetic analysis of rotavirus (RV) strains based on VP3 gene sequences; (**C**) phylogenetic analysis of canine parvovirus (CPV-2) strains based on VP2 gene sequences. The sequences obtained in the present study are marked in red.

4. Discussion

Ever since it was first developed and optimized in the early 1980s [19,20], the PCR test has been a powerful tool in molecular biology [21,22]. The utilization of PCR in different ways permits the rapid and efficient recognition of DNA or RNA sequences of various origins, and benefits relevant fields of both fundamental research and practical application. Here, in the current study, we established a multiplex PCR (mPCR) method that allows the detection of three viruses, either DNA- or RNA-based, simultaneously in giant pandas. The results indicate that the mPCR method is steadfast, with matter-of-fact outcomes from clinical fecal samples that can easily be interpreted.

Conventionally, the mPCR method is considered to have relatively reduced sensitivity and specificity, as multiple PCR reactions occur at the same time, with an inherent risk of competing for polymerases within reactions, and because of the cross-reactivity of the primer pairs in use. Nevertheless, through the development of bioinformatic databases and reagents for molecular biology research, the reliability of mPCR can be largely improved in both areas [23]. The careful design of primers, including the use of homology analysis against pathogens and host sequences, could efficiently avoid the occurrence of cross-reactions or nonspecific reactions with other pathogens or host genetic materials [24]. In the meantime, the fast development of commercial polymerases also enables the robust competence of the PCR reaction, even if multiple primers are involved. In the present study, the established mPCR method ran smoothly across all clinical samples, with clear indications of positive or negative outcomes without the ambiguous amplification that results in unexplainable PCR bands. Meanwhile, the sensitivity of the results was 1×10^1 (CDV) or 1×10^2 (CPV-2 and GPRV) viral copies, which was already comparable to that of the traditional PCR. As such, our observations indicated that the established mPCR technique is reliable and can be reinforced in routine molecular diagnostics of viral infections in giant pandas.

The fast development of viromics research and bioinformatics techniques benefited the field of epidemiology greatly, where viral host spectrums were identified via advanced sequencing [25,26]. Newly identified pathogens and their associated diseases demand effective molecular diagnostic methods, especially ones accessible to laboratories with fewer facilitaties. As a conventional technique, the mPCR method has the inherent advantage of being cost-effective, easily operated, and less time-consuming [27–29]. Therefore, mPCR at present could serve as a part of a fast-reacting system that allows for rapid detection under circumstances such as an acute outbreak and provides first-hand etiological information that could assist clinical treatment or epidemiological policy making.

The cross-species transmission of various pathogens poses a risk to endangered species such as giant pandas. Contact with humans or other animals, albeit often well protected, also leads to an inevitable consequence of reverse zoonosis or horizonal transmission, via which bacterial or viral infectious agents are introduced to the giant panda population [30–32]. Utilizing the established mPCR method, we showed that CDV, GPRV, and CPV-2 were detected in giant pandas, and that co-infection with two or three viruses was commonly observed. The observation was further consolidated with phylogenetic analysis of the matching full-length sequences of mPCR targets. No significant correlation was spotted between symptoms of sampled animals and virus detection in the present study, indicating that infection with these viruses was mostly subclinical, and that acute phases of infection might be neglected or wrongly interpreted. Among the three viruses evaluated, CPV-2 was the most frequently (19.72%) detected in Giant pandas during this investigation. CPV-2 and related viruses of species *Protoparvovirus Carnivoran 1* can infect a wide range of domestic and wild carnivores, which leads to their worldwide distribution with a broad host spectrum [33,34]. Noticeably, due to the close genetic relationship between CPV and feline panleukopenia virus (FPV) in their VP2 sequences and the entire genome [35,36], our method (or any approach without DNA sequencing) cannot differentiate between infection with CPV-2 and FPV, the latter of which was recently reported as being able to infect giant pandas [37]. Further analysis of the full-length VP2 sequences indicated that all amplified parvoviral sequences belong to CPV-2, showing that CPV-2 is the main component of *Protoparvovirus Carnivoran 1* in giant pandas, but the potential threats of FPV transmission in giant pandas should not be neglected. Meanwhile, a smaller but unneglectable number of samples are positive for CDV or GPRV, or both. To this point, recurrent epidemiological surveillance is essential to monitor possible increased transmission of the three virus's infection in giant pandas and possibly other wild animal species.

In conclusion, we established and validated an mPCR method that allows the rapid and accurate detection of CPV-2, CDV, and GPRV infection. The complex nature of spatial viral infection in giant pandas was revealed via this method, and the key observations justify further investigation on the infection mechanisms of those viruses from both epidemiological and etiological perspectives.

Supplementary Materials: The following supporting information can be downloaded at: https://www.mdpi.com/article/10.3390/vetsci12020081/s1, Figure S1: Detection of PCR electrophoretograms of giant panda samples by mPCR.; Table S1: Viruses able to infect giant panda [38–51].

Author Contributions: Conceptualization, Y.L. and S.Z.; methodology, Y.L.; software, R.M.; validation, A.L., and W.Q.; formal analysis, A.L.; investigation, A.L.; resources, W.Q.; data curation, R.M.; writing—original draft preparation, A.L.; writing—review and editing, Y.L.; visualization, W.Q.; supervision, S.Z. and Q.Y.; project administration, Y.L.; funding acquisition, Y.L. All authors have read and agreed to the published version of the manuscript.

Funding: This work was supported by the Joint Funds of Science Technology and Education of Sichuan Province (2024NSFSC2063), and Local Financial Funds of National Agricultural Science and Technology Center, Chengdu (NASC2023ST11).

Institutional Review Board Statement: Not applicable.

Informed Consent Statement: Not applicable.

Data Availability Statement: Data are contained within the article and Supplementary Material.

Conflicts of Interest: The authors declare no conflicts of interest.

References

1. Simmonds, P.; Adams, M.J.; Benko, M.; Breitbart, M.; Brister, J.R.; Carstens, E.B.; Davison, A.J.; Delwart, E.; Gorbalenya, A.E.; Harrach, B.; et al. Consensus statement: Virus taxonomy in the age of metagenomics. *Nat. Rev. Microbiol.* **2017**, *15*, 161–168. [CrossRef] [PubMed]
2. Goodwin, C.M.; Xu, S.; Munger, J. Stealing the Keys to the Kitchen: Viral Manipulation of the Host Cell Metabolic Network. *Trends Microbiol.* **2015**, *23*, 789–798. [CrossRef] [PubMed]
3. Walsh, D.; Mohr, I. Viral subversion of the host protein synthesis machinery. *Nat. Rev. Microbiol.* **2011**, *9*, 860–875. [CrossRef] [PubMed]
4. Hao, P.; Pang, Z.; Qu, Q.; Cui, C.; Jiang, Y.; Chen, J.; Gao, Z.; Xu, Z.; Li, L.; Jin, N.; et al. A G3P[3] bat rotavirus can infect cultured human cholangiocytes and cause biliary atresia symptom in suckling mice. *Virol. Sin.* **2024**, *39*, 974–976. [CrossRef]
5. Li, J.; Peng, J.; Zeng, Y.; Wang, Y.; Li, L.; Cao, Y.; Cao, L.; Chen, Q.; Ye, Z.; Zhou, D.; et al. Isolation of a feline-derived feline panleukopenia virus with an A300P substitution in the VP2 protein and confirmation of its pathogenicity in dogs. *Anim. Dis.* **2024**, *4*, 5. [CrossRef]
6. Zhou, Z.; Qiu, Y.; Ge, X. The taxonomy, host range and pathogenicity of coronaviruses and other viruses in the Nidovirales order. *Anim. Dis.* **2021**, *1*, 5. [CrossRef]
7. Parrish, C.R.; Holmes, E.C.; Morens, D.M.; Park, E.C.; Burke, D.S.; Calisher, C.H.; Laughlin, C.A.; Saif, L.J.; Daszak, P. Cross-species virus transmission and the emergence of new epidemic diseases. *Microbiol. Mol. Biol. Rev.* **2008**, *72*, 457–470. [CrossRef]
8. Dobson, A.; Foufopoulos, J. Emerging infectious pathogens of wildlife. *Philos. Trans. R. Soc. Lond. B Biol. Sci.* **2001**, *356*, 1001–1012. [CrossRef] [PubMed]
9. Zhang, W.; Yang, S.; Shan, T.; Hou, R.; Liu, Z.; Li, W.; Guo, L.; Wang, Y.; Chen, P.; Wang, X.; et al. Virome comparisons in wild-diseased and healthy captive giant pandas. *Microbiome* **2017**, *5*, 90. [CrossRef] [PubMed]
10. Ning, S.; Lu, X.; Zhao, M.; Wang, X.; Yang, S.; Shen, Q.; Wang, H.; Zhang, W. Virome in Fecal Samples From Wild Giant Pandas (*Ailuropoda melanoleuca*). *Front. Vet. Sci.* **2021**, *8*, 767494. [CrossRef] [PubMed]
11. Shi, X.; Song, G.; Li, Z. The effect of diffusion on giant pandas that live in complex patchy environments. *Nonlinear Anal. Model. Control.* **2015**, *20*, 56–71. [CrossRef]

12. Ran, M.X.; Li, Y.; Zhang, Y.; Liang, K.; Ren, Y.N.; Zhang, M.; Zhou, G.B.; Zhou, Y.M.; Wu, K.; Wang, C.D.; et al. Transcriptome Sequencing Reveals the Differentially Expressed lncRNAs and mRNAs Involved in Cryoinjuries in Frozen-Thawed Giant Panda (*Ailuropoda melanoleuca*) Sperm. *Int. J. Mol. Sci.* **2018**, *19*, 3066. [CrossRef] [PubMed]
13. Wang, T.; Xie, Y.; Zheng, Y.; Wang, C.; Li, D.; Koehler, A.V.; Gasser, R.B. Parasites of the Giant Panda: A Risk Factor in the Conservation of a Species. *Adv. Parasitol.* **2018**, *99*, 1–33. [CrossRef] [PubMed]
14. Mainka, S.A.; Qiu, X.; He, T.; Appel, M.J. Serologic survey of giant pandas (*Ailuropoda melanoleuca*), and domestic dogs and cats in the Wolong Reserve, China. *J. Wildl. Dis.* **1994**, *30*, 86–89. [CrossRef]
15. Qin, Q.; Li, D.; Zhang, H.; Hou, R.; Zhang, Z.; Zhang, C.; Zhang, J.; Wei, F. Serosurvey of selected viruses in captive giant pandas (*Ailuropoda melanoleuca*) in China. *Vet. Microbiol.* **2010**, *142*, 199–204. [CrossRef] [PubMed]
16. Guo, L.; Yan, Q.; Yang, S.; Wang, C.; Chen, S.; Yang, X.; Hou, R.; Quan, Z.; Hao, Z. Full Genome Sequence of Giant Panda Rotavirus Strain CH-1. *Genome Announc.* **2013**, *1*, e00241-12. [CrossRef] [PubMed]
17. Li, D.; Zhu, L.; Cui, H.; Ling, S.; Fan, S.; Yu, Z.; Zhou, Y.; Wang, T.; Qian, J.; Xia, X.; et al. Influenza A(H1N1)pdm09 virus infection in giant pandas, China. *Emerg. Infect. Dis.* **2014**, *20*, 480–483. [CrossRef] [PubMed]
18. Zhao, M.; Yue, C.; Yang, Z.; Li, Y.; Zhang, D.; Zhang, J.; Yang, S.; Shen, Q.; Su, X.; Qi, D.; et al. Viral metagenomics unveiled extensive communications of viruses within giant pandas and their associated organisms in the same ecosystem. *Sci. Total Environ.* **2022**, *820*, 153317. [CrossRef]
19. Saiki, R.K.; Gelfand, D.H.; Stoffel, S.; Scharf, S.J.; Higuchi, R.; Horn, G.T.; Mullis, K.B.; Erlich, H.A. Primer-directed enzymatic amplification of DNA with a thermostable DNA polymerase. *Science* **1988**, *239*, 487–491. [CrossRef]
20. Sanger, F.; Coulson, A.R. A rapid method for determining sequences in DNA by primed synthesis with DNA polymerase. *J. Mol. Biol.* **1975**, *94*, 441–448. [CrossRef] [PubMed]
21. McDowell, D. The polymerase chain reaction patents: Going, going,... still going. *J. R. Soc. Med.* **2006**, *99*, 62–64. [CrossRef] [PubMed]
22. Cavé, H. The polymerase chain reaction, so simple, so clever: The discovery that made minimal residual disease come true. *Haematologica* **2022**, *107*, 1737–1738. [CrossRef]
23. Elnifro, E.M.; Ashshi, A.M.; Cooper, R.J.; Klapper, P.E. Multiplex PCR: Optimization and application in diagnostic virology. *Clin. Microbiol. Rev.* **2000**, *13*, 559–570. [CrossRef] [PubMed]
24. Brownie, J.; Shawcross, S.; Theaker, J.; Whitcombe, D.; Ferrie, R.; Newton, C.; Little, S. The elimination of primer-dimer accumulation in PCR. *Nucleic Acids Res.* **1997**, *25*, 3235–3241. [CrossRef]
25. Schulz, F.; Roux, S.; Paez-Espino, D.; Jungbluth, S.; Walsh, D.A.; Denef, V.J.; McMahon, K.D.; Konstantinidis, K.T.; Eloe-Fadrosh, E.A.; Kyrpides, N.C.; et al. Giant virus diversity and host interactions through global metagenomics. *Nature* **2020**, *578*, 432–436. [CrossRef]
26. Visser, M.; Bester, R.; Burger, J.T.; Maree, H.J. Next-generation sequencing for virus detection: Covering all the bases. *Virol. J.* **2016**, *13*, 85. [CrossRef] [PubMed]
27. Rajkhowa, S.; Choudhury, M.; Pegu, S.R.; Sarma, D.K.; Gupta, V.K. Development of a novel one-step triplex PCR assay for the simultaneous detection of porcine circovirus type 2, porcine parvovirus and classical swine fever virus in a single tube. *Lett. Appl. Microbiol.* **2022**, *75*, 338–344. [CrossRef]
28. Hao, X.; Liu, R.; He, Y.; Xiao, X.; Xiao, W.; Zheng, Q.; Lin, X.; Tao, P.; Zhou, P.; Li, S. Multiplex PCR methods for detection of several viruses associated with canine respiratory and enteric diseases. *PLoS ONE* **2019**, *14*, e0213295. [CrossRef]
29. Xiao, X.; Hao, X.; Chen, B.; Zhou, P.; Li, S. Two Multiplex PCR Methods for Detecting Several Pathogens Associated with Feline Respiratory and Intestinal Tracts. *Vet. Sci.* **2022**, *10*, 14. [CrossRef] [PubMed]
30. Mertz, G.J. Zoonoses: Infectious Diseases Transmissible From Animals to Humans, Fourth Edition. *Clin. Infect. Dis.* **2011**, *63*, 148–149. [CrossRef]
31. Allen, T.; Murray, K.A.; Zambrana-Torrelio, C.; Morse, S.S.; Rondinini, C.; Di Marco, M.; Breit, N.; Olival, K.J.; Daszak, P. Global hotspots and correlates of emerging zoonotic diseases. *Nat. Commun.* **2017**, *8*, 1124. [CrossRef]
32. Allocati, N.; Petrucci, A.G.; Di Giovanni, P.; Masulli, M.; Di Ilio, C.; De Laurenzi, V. Bat-man disease transmission: Zoonotic pathogens from wildlife reservoirs to human populations. *Cell Death Discov.* **2016**, *2*, 16048. [CrossRef] [PubMed]
33. Ndiana, L.A.; Lanave, G.; Desario, C.; Berjaoui, S.; Alfano, F.; Puglia, I.; Fusco, G.; Colaianni, M.L.; Vincifori, G.; Camarda, A.; et al. Circulation of diverse protoparvoviruses in wild carnivores, Italy. *Transbound. Emerg. Dis.* **2021**, *68*, 2489–2502. [CrossRef] [PubMed]
34. Qi, S.; Zhao, J.; Guo, D.; Sun, D. A Mini-Review on the Epidemiology of Canine Parvovirus in China. *Front. Vet. Sci.* **2020**, *7*, 5. [CrossRef] [PubMed]
35. Leopardi, S.; Milani, A.; Cocchi, M.; Bregoli, M.; Schivo, A.; Leardini, S.; Festa, F.; Pastori, A.; de Zan, G.; Gobbo, F.; et al. Carnivore protoparvovirus 1 (CPV-2 and FPV) Circulating in Wild Carnivores and in Puppies Illegally Imported into North-Eastern Italy. *Viruses* **2022**, *14*, 2612. [CrossRef] [PubMed]

36. Calatayud, O.; Esperón, F.; Velarde, R.; Oleaga, Á.; Llaneza, L.; Ribas, A.; Negre, N.; de la Torre, A.; Rodríguez, A.; Millán, J. Genetic characterization of Carnivore Parvoviruses in Spanish wildlife reveals domestic dog and cat-related sequences. *Transbound. Emerg. Dis.* **2020**, *67*, 626–634. [CrossRef] [PubMed]
37. Yi, S.; Liu, S.; Meng, X.; Huang, P.; Cao, Z.; Jin, H.; Wang, J.; Hu, G.; Lan, J.; Zhang, D.; et al. Feline Panleukopenia Virus With G299E Substitution in the VP2 Protein First Identified From a Captive Giant Panda in China. *Front. Cell. Infect. Microbiol.* **2022**, *11*, 820144. [CrossRef] [PubMed]
38. Guo, L.; Yang, S.L.; Chen, S.J.; Zhang, Z.; Wang, C.; Hou, R.; Ren, Y.; Wen, X.; Cao, S.; Guo, W.; et al. Identification of canine parvovirus with the Q370R point mutation in the VP2 gene from a giant panda (*Ailuropoda melanoleuca*). *Virol. J.* **2013**, *10*, 163. [CrossRef] [PubMed]
39. Dai, Z.; Wang, H.; Feng, Z.; Ma, L.; Yang, S.; Shen, Q.; Wang, X.; Zhou, T.; Zhang, W. Identification of a novel circovirus in blood sample of giant pandas (*Ailuropoda melanoleuca*). *Infect. Genet. Evol.* **2021**, *95*, 105077. [CrossRef]
40. Qi, D.; Shan, T.; Liu, Z.; Deng, X.; Zhang, Z.; Bi, W.; Owens, J.R.; Feng, F.; Zheng, L.; Huang, F.; et al. A novel polyomavirus from the nasal cavity of a giant panda (*Ailuropoda melanoleuca*). *Virol. J.* **2017**, *14*, 207. [CrossRef] [PubMed]
41. Yang, R.; Wang, C.; Yan, Q. Research progress of giant panda rotavirus strain CH-1. *Chin. J. Zoonoses* **2018**, *34*, 1040–1043. (In Chinese)
42. Su, X.; Li, L.; Yan, X.; Zhang, D.; Hou, R.; Liu, S. Establishment and application of giant panda rotavirus PCR detection method. *Acta Theriol. Sin.* **2021**, *41*, 254–260. (In Chinese)
43. Wang, C.; Yan, Q.; Zhang, Z.; Luo, L.; Fan, W.; Yang, Z.; Lan, J.; Huang, X.; Li, M. Isolation and identification of rotavirus from giant panda cubs. *Acta Theriol. Sin.* **2008**, *1*, 87–91. (In Chinese)
44. Gao, F.S.; Hu, G.X.; Xia, X.Z.; Gao, Y.W.; Bai, Y.D.; Zou, X.H. Isolation and identification of a canine coronavirus strain from giant pandas (*Ailuropoda melanoleuca*). *J. Vet. Sci.* **2009**, *10*, 261–263. [CrossRef] [PubMed]
45. Qiao, J.; Xia, X.; Yang, S.; Li, D.; Hu, G.; Gao, Y.; Sun, H.; Zhao, Z.; Xie, Z.; Yan, F.; et al. Serological survey on canine coronavirus antibodies in giant pandas by virus neutralization test. *J. For. Res.* **2004**, *15*, 295–297.
46. Zhang, C.; Ding, Y.; Yan, H.; Pu, T.; She, R.; Yin, J.; Yang, M.; Zhang, J. Survey of Hepatitis E Virus Infection in Giant Panda. *Chin. J. Wildl.* **2013**, 323–326. (In Chinese) [CrossRef]
47. Feng, N.; Yu, Y.; Wang, T.; Wilker, P.; Wang, J.; Li, Y.; Sun, Z.; Gao, Y.; Xia, X. Fatal canine distemper virus infection of giant pandas in China. *Sci. Rep.* **2016**, *6*, 27518. [CrossRef] [PubMed]
48. Zhao, N.; Li, M.; Luo, J.; Wang, S.; Liu, S.; Wang, S.; Lyu, W.; Chen, L.; Su, W.; Ding, H.; et al. Impacts of canine distemper virus infection on the giant panda population from the perspective of gut microbiota. *Sci. Rep.* **2017**, *7*, 39954. [CrossRef]
49. Geng, Y.; Shen, F.; Wu, W.; Zhang, L.; Luo, L.; Fan, Z.; Hou, R.; Yue, B.; Zhang, X. First demonstration of giant panda's immune response to canine distemper vaccine. *Dev. Comp. Immunol.* **2020**, *102*, 103489. [CrossRef]
50. Hvistendahl, M. Captive pandas succumb to killer virus. *Science* **2015**, *347*, 700–701. [CrossRef]
51. Guo, L.; Li, Z.; Yang, S. Primary research and establishment of Rabies virus RT-PCR detection method. *Proc. Annu. Conf. Sichuan Anim. Husb. Vet. Soc.* **2012**, 387–391. (In Chinese)

veterinary sciences

MDPI

Review

Current Insights into Porcine Bocavirus (PBoV) and Its Impact on the Economy and Public Health

Jelena Prpić [1,*], Tomislav Keros [2], Margarita Božiković [1], Magda Kamber [1] and Lorena Jemeršić [1]

1 Croatian Veterinary Institute, Savska Cesta 143, 10000 Zagreb, Croatia
2 Sanatio d.o.o., Nehajska 10, 10000 Zagreb, Croatia
* Correspondence: balatinec@veinst.hr

Simple Summary: Porcine Bocavirus (PBoV) is an emerging viral pathogen in swine that has not received significant attention despite its potential impact. This review aims to fill the knowledge gap by analyzing the existing literature on PBoV, including its genome structure, discovery, classification, detection methods, and pathogenesis. This review also explores the public health implications and economic impact of PBoV, such as decreased productivity, increased veterinary costs, and trade restrictions. By enhancing understanding of PBoV, this review seeks to aid in its prevention and control, thereby mitigating its economic impact on the swine industry.

Abstract: Effective control of animal infectious diseases is crucial for maintaining robust livestock production systems worldwide. Porcine meat constitutes approximately 35–40% of global meat production with the largest producers being China and the European Union (EU). Emerging viral pathogens in swine, like porcine bocavirus (PBoV), have not garnered significant attention, leaving their pathogenic characteristics largely unexplored. This review aims to bridge this knowledge gap by conducting a comprehensive analysis of the existing literature on PBoV. We explore the virus's genome structure, discovery, classification, detection methods, pathogenesis, and its potential public health implications. Additionally, we discuss the distribution and economic impact of PBoV, which includes potential losses due to decreased productivity, increased veterinary costs, and trade restrictions. By highlighting the current state of knowledge, this review seeks to enhance the understanding of PBoV, thereby aiding in its prevention and control, and mitigating its economic impact on the swine industry.

Keywords: porcine bocavirus (PBoV); genome structure; classification; detection; public health concerns; economic impact

1. Introduction

Pig meat is a major source of protein in the human diet, with a stable share of 35–40% of global meat production [1]. In Europe, the pig meat sector accounts for about 150 million reared pigs, representing nearly half of the total EU meat production [2]. The EU is the world's second biggest producer of pork after China and the biggest exporter of pork and pork products, especially after the reduction in pork production in Asia, caused by African Swine Fever (ASF) [3]. However, even in countries of insufficient pig production, such as Croatia, high import can also be a high risk for disease introduction with a potentially severe economic impact. In Croatia, the members of the Croatian Association of Pork Producers are the largest producers of piglets, fattening pigs, and pig genetic material with a yearly production of about 700,000 fattening pigs and 30,000 sows [4], and according to the number of pigs and pig products, the Croatian contribution within the EU is below 1% [5]. However, 70 percent of Croatia's pork meat market is covered by imports from other EU countries mostly from Germany, Spain, Hungary, the Netherlands, and Denmark [6] and is the 32nd largest importer of pig meat in the world.

Infectious agents, such as bacteria and viruses, are the main threats to production in intensive pig farms. Historically, many viral diseases, such as Porcine Reproductive and Respiratory Syndrome (PRRS), Porcine Circovirus-associated Diseases (PCVADs), and African Swine Fever and Foot and Mouth Disease (FMD), had a negative impact on the swine industry causing great economic losses [3,7–11]. Moreover, the zoonotic potential of viruses such as the causative agents of highly pathogenic avian influenza virus (HPAI), Severe Acute Respiratory Syndrome (SARS), and Middle East Respiratory Syndrome (MERS) have been shown to be of high risk for human health and should be identified in a timely manner. Recently, a number of small novel porcine DNA viruses have emerged globally, for example, torque teno sus virus, porcine bocavirus (PBoV), and parvoviruses. However, there is still a lack of knowledge regarding their biology, interspecies transmission, and potential pathogenicity. A newly described genetically unique class of viruses known as bocaparvovirus has been found in both humans and animals. One of the representatives of bocaviruses (BoVs) is an emerging pathogen that was recognized to cause a great negative impact on swine industry, namely PBoV. PBoV has currently been reported worldwide, mostly in weaning piglets. Since PBoV is common in both clinically and healthfully infected pigs, it is typically linked to co-infection with other viruses [12,13]. Therefore, it is important to understand PBoV and its pathogenic nature. Additionally, PBoV presents a serious risk to public health. A case study from Iran suggested that since PBoV was found in a three-year-old child who had an acute respiratory tract infection, the virus may be evolving as a human pathogen [14].

A study regarding PBoV prevalence in Croatia [15] reported its circulation among the Croatian domestic pig population. A conducted survey showed that PBoV is circulating among Croatian intensive pig farms and small family farms, and that PBoV might have been introduced by imported pigs. In order to protect a thriving pig industry and to facilitate early detection of the PoBV infection and its prevention and control, we aimed to summarize the current knowledge about PBoV, including the discovery, classification, genome structure, main diagnostic methods, pathogenesis, public health importance, distribution, and economic impact of PBoV.

2. Discovery and Classification

Although BoVs were recognized in the early 1960s, PBoV was not isolated until 2009. Initially referred to as porcine boca-like virus (PBo-likeV), it was first detected in the lymph nodes of Swedish pigs suffering from postweaning multisystemic wasting syndrome (PMWS) [16,17]. In 2010, PBo-likeV was discovered in China and officially named PBoV or PBoV1 [18,19]. Since its discovery, PBoV has been reported globally (Figure 1; Table 1), including in Europe [15,20–26], Asia [27–30], North America [31,32], and Africa [33,34]. In recent studies, the detection of porcine bocavirus (PBoV) has been reported in various countries, highlighting its widespread presence and potential impact on swine health. Table 1 provides a comprehensive overview of the detection rates of PBoV in different countries, along with key details on the age of affected pigs and the sources of the samples collected. This information is crucial for understanding the epidemiology of PBoV and identifying the most vulnerable populations. Globally, the prevalence rate of PBoV was 46% in pigs without PMWS and 88% in pigs with PMWS [35]. In China, the prevalence rate of PBoV in stool samples of piglets was found to be 12.59%. Similarly, the prevalence rate of bocaviruses in clinical samples from one-month-old piglets was reported to be 5.77%. In the USA, studies have documented prevalence rates in sick pigs of approximately 43% and 59%. PBoV has been identified in various regions, including Africa. However, specific prevalence rates in Africa are not detailed in the sources. The virus is prevalent in both healthy and clinically infected pigs, often found in weaning piglets and associated with coinfections with other viruses [35]. The virus's presence in these regions highlights its widespread nature and the potential for significant impacts on pig health and the swine industry across the continent. In Croatia, PBoV has been identified in pigs, contributing to the understanding of its epidemiology in Eastern Europe [15,35]. Studies have shown that the virus is present in both healthy and diseased pigs, indicating its pervasive nature and the need for ongoing surveillance and research to mitigate its impact on the swine industry [15,35].

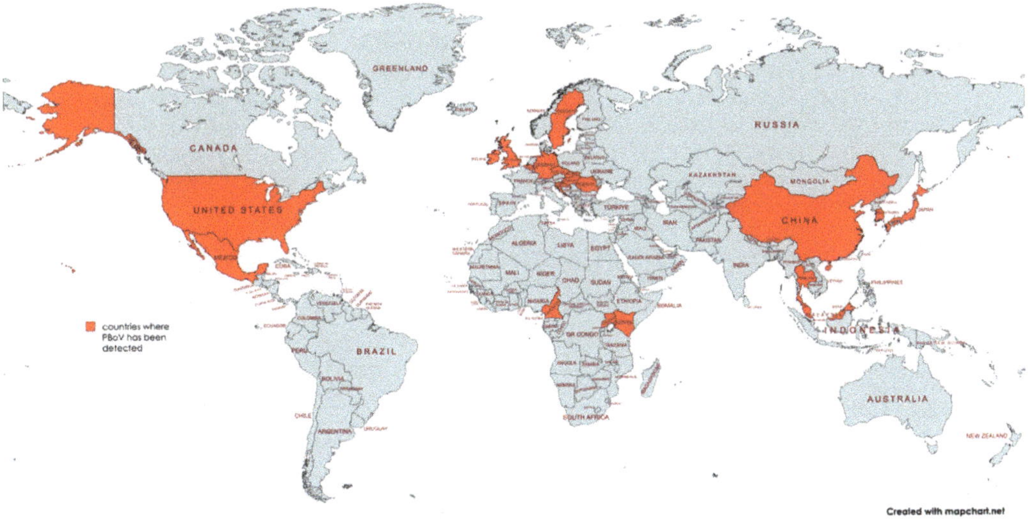

Figure 1. Geographical distribution of PBoV. Map is created with Mapchart [36].

Table 1. Detection of PBoV in various countries: key details on age of affected pigs and sample sources.

Country	Year of Identification	Sample Source	Age Categories	References
Sweden	2009	Lymph nodes	weaners	[16,17]
China	2010	Fecal samples	weaners	[18]
USA	2010	Lung, lymph nodes, spleen	unknown	[31]
Ireland	2011	Spleen	piglets, weaners, sows	[21]
Romania	2011	Lymph nodes, lung, kidney, liver, spleen, tonsil	piglets, yearlings, adults wild boar	[20]
Hungary	2012	Fecal samples, blood serum samples, organ tissues, fetuses, semen	weaners	[22]
Croatia	2013	Fecal samples	fatteners	[15]
Cameroon	2013	Fecal samples	piglets	[33]
UK	2014	Lung, liver, kidney, spleen, lymph nodes, serum samples	unknown	[26]
Thailand	2014	Tonsil	rearing pigs	[27]
Korea	2014	Serum and fecal samples, saliva	weaners	[28]
Czech Republic	2014	Lymph nodes, liver, spleen	piglets, weaners, fatteners	[23]
Slovakia	2014	Lymph nodes, liver, spleen	piglets, weaners, fatteners	[23]
Mexico	2015	Serum samples	unknown	[32]
Germany	2016	Lung, lymph nodes	weaners	[24]
Japan	2017	Tonsil	unknown	[29]
Uganda	2017	Fecal samples	unknown	[34]
Kenya	2017	Fecal samples	unknown	[34]
Slovenia	2017	Fecal samples	unknown	[15]
Belgium	2018	Fecal samples	fatteners	[25]
Malaysia	2018	Mesenteric lymph node, submandibular lymph node, inguinal lymph node, spleen, tonsil, lung, kidney, liver	piglets, weaners	[30]

BoVs belong to the genera *Bocaparvovirus*, family *Parvoviridae*, and subfamily *Parvovirinae* [37,38]. In contrast to other parvoviruses, BoVs contain a third open reading frame in the middle of the genome encoding for a highly phosphorylated non-structural protein, NP1, whose function has not yet been determined.

Their hosts are, as known to date, different mammalian species including primates (Table 2). Furthermore, the classification of PBoV within the broader context of the Bocaparvovirus genus is essential for understanding its relationship with other viruses in the Parvovirinae subfamily. Table 2 summarizes the viral species belonging to the genus Bocaparvovirus, as suggested by the International Committee on Taxonomy of Viruses (ICTV). This classification helps in elucidating the genetic and evolutionary relationships between PBoV and other related viruses, providing insights into their pathogenesis and potential cross-species transmission.

Table 2. Summary of viral species belonging to the genus *Bocaparvovirus*, subfamily *Parvovirinae*, and family *Parvoviridae*, as suggested by the ICTV [26].

Host	Species	Virus Name	Abbreviation	References
family *Canidae*	*Carnivore bocaparvovirus* 1	Canine minute virus; minute virus of canines	CnMV; MVC	[39,40]
	Carnivore bocaparvovirus 2	Canine bocavirus 2	CBoV2	[41]
	Carnivore bocaparvovirus 7	Canine bocavirus 3	CBoV3	[42]
family *Felidae*	*Carnivore bocaparvovirus* 3	Feline bocavirus 1	FBoV1	[43]
	Carnivore bocaparvovirus 4	Feline bocaparvovirus 2	FboV2	[44]
	Carnivore bocaparvovirus 5	Feline bocaparvovirus 3	FBoV3	[45]
family *Mustelidae*	*Carnivore bocaparvovirus* 6	Mink bocavirus 1	MiBoV1	[46]
family *Vespertilionidae*	*Chiopteran bocaparvovirus* 1	Myotis myotis (bat) bocavirus 1	BtBoV1	[47]
	Chiopteran bocaparvovirus 2	Bat bocavirus WM40	BtBoVwM40	[48]
	Chiopteran bocaparvovirus 3	Bat bocavirus XM30	BtBoVxm30	[49]
	Chiopteran bocaparvovirus 4	Miniopterus schreibersii bat bocavirus	BtBoV2	[40]
	Chiopteran bocaparvovirus 5	Rousettus leschenaultia bocaparvovirus 1	RIBoV	[50]
family *Leporidae*	*Lagomorph bocaparvovirus* 1	Rabbit bocaparvovirus	RBoV	[51]
family *Otariidae*	*Pinniped bocaparvovirus* 1	California sea lion bocavirus 1	CsIBoV1	[40]
	Pinniped bocaparvovirus 2	California sea lion bocavirus 3	CsIBoV3	[40]
family *Hominidae*	*Primate bocaparvovirus* 1	Human bocavirus 1 and 3	HBoV1, 3	[52]
	Primate bocaparvovirus 2	Human bocavirus 2 and 4	HBoV2, 4	[53]
family *Cercopithecidae*	*Primate bocaparvovirus* 3	Macca mulatta bocaparvovirus	MmBoV	[54]
family *Muridae*	*Rodent bocaparvovirus* 1	Rat bocavirus	RBoV	[50]
	Rodent bocaparvovirus 2	Murine bocavirus	MuBoV	[55]
family *Bovidae*	*Ungulate bocaparvovirus* 1	Bovine parvovirus 1	BPV1	[56]
	Ungulate bocaparvovirus 6	Bovine bocaparvovirus 2	BBoV2	[57]
family *Suidae*	*Ungulate bocaparvovirus* 2	Porcine bocavirus 1	PBoV1	[16]
	Ungulate bocaparvovirus 3	Porcine bocavirus SX	PBoVsx	[58]
	Ungulate bocaparvovirus 4	Porcine bocavirus H18	PBoVh18	[59]
	Ungulate bocaparvovirus 5	Porcine bocavirus 3	PBoV3	[60]
family *Camelidae*	*Ungulate bocaparvovirus* 7	Dromedary camel bocaparvovirus 1	DBoV1	[61]
	Ungulate bocaparvovirus 8	Dromedary camel bocaparvovirus 2	DBoV2	[61]
	Ungulate bocaparvovirus 9	Vicugna pacos bocaparvovirus	VpBoV	[62]

BoV strains that have been isolated from different hosts are closely related to PBoV. For instance, strains isolated from mink share 87% nucleotide similarity with PBoV (strain: PBoV-KU14; HQ223038, HQ291308, and KJ622366) and form a distinct clade within the PBoV clades [46,63]. Similarly, strains such as those isolated from bats [48,49,64], rodents [50,55,65], and Himalayan marmots [54] are also closely related to PBoV. These studies suggest that rodents, minks, bats, Himalayan marmots, and other species may have transmitted diseases to pigs. To date, however, there has been no reported experimental evidence of BoVs' interspecies transmission. These investigations demonstrate the great genetic diversity and broad host adaptation of BoVs. Figure 2 illustrates the phylogenetic connections between porcine bocavirus (PBoV) and other bocaviruses.

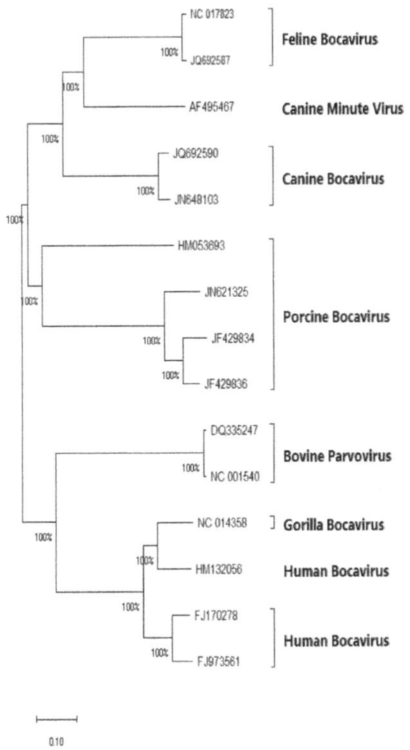

Figure 2. Phylogenetic tree of PBoV and related viruses based on the complete genome sequences. The tree was constructed using the Neighbor-Joining method with 1000 bootstrap replicates. Bootstrap values greater than 50% are shown at the branch nodes. The tree includes various BoV strains from different hosts, including humans, pigs, dogs, cats, cows, and gorillas. The scale bar represents the number of nucleotide substitutions per site. The phylogenetic tree reveals the relationships between PBoV and other bocaviruses. Notably, PBoV is closely related to other bocaviruses, including feline bocavirus and bovine bocavirus.

3. Genome Structure

PBoV is a nonenveloped single-stranded DNA virus which exhibits icosahedral symmetry [12,19] and has a diameter of 26–30 nm [66]. It is a linear, single-stranded DNA of 4–6 kb [12,19] and can have either positive or, more frequently, negative polarity [67] and contains three open reading frames (ORFs): ORF1, ORF2, and ORF3, respectively [53,68]. Specifically: ORF1 encodes a nonstructural protein 1 (NS1) which encodes a protein involved in ATPase and helicase activity and rolling circle replication [68]. Bovine parvovirus (BPV) and PBoV's ORF-1 region are similar [60]. ORF2 encodes viral capsid proteins 1 and 2 (VP1/2), where VP1 includes the entire VP2 sequence and an additional N-terminal region [69]. ORF-2 encodes a protein linked to catalytic residues (the most conserved motif HDXXY), calcium-binding loop, and phospholipase A2 motifs (the most conserved motif YXGXG) necessary for parvovirus infectivity [12,19]. PBoV's conserved motif, YXGXF, is distinct from other parvoviruses' YXGXG motif [12,19]. The ORF-2 region of PBoV resembles canine minute virus (CMV) [61]. ORF3 encodes nuclear phosphoprotein 1 (NP1) of unclear function (Figure 3) [60].

The classification of PBoV into genotypes is based on nucleotide and amino acid alignments. Specifically, PBoV strains are grouped into different clades according to the sequences of the VP1 and VP2 genes [70]. The nucleotide sequence identity cutoff for distinguishing genotypes is generally between 78% and 81% [56]. Based on the VP1 and VP2 sequences, PBoV has been classified into different clades, including PBoV1, PBoV2, PBoV3, PBoV4, PBoV5, PBoV3C, PBoV-6V, and PBoV-7V. Additionally, PBoV has been proposed to be classified into three different groups: PBoV G1, PBoV G2, and PBoV G3 [71].

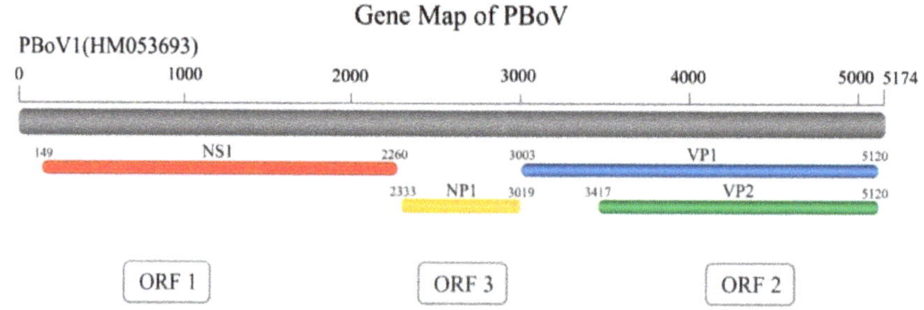

Figure 3. The genome structure of PBoV (adapted from [71]). The PBoV genome of approximately 5.3 bb contains three ORFs that encode for four proteins (NS1, NP1, VP1, and VP2).

PBoV shows significant diversity, prevalence, and complexity among its strains. Phylogenetic and recombination analyses have revealed considerable variation in the genome sequences of different PBoV strains. For example, the NS1 gene of PBoV shares sequence identities with other bocaviruses, such as CMV, BPV, and human bocavirus (HBoV), ranging from 30% to 54.7% [71].

4. Main Diagnostic Methods

There are several PCR-based methods for rapid PBoV detection. For detection of the three groups of PBoV, conventional multiplex PCR can be successfully used. Multiplex PCR amplifies multiple DNA regions simultaneously using different sets of primers in a single PCR reaction. Single PCR assays and the multiplex PCR assay have equal sensitivity of 1.0×10^3 genomic copies/lL for PBoV group 1, 4.5×10^3 copies/lL for PBoV group 2, and 3.8×10^3 copies/lL for PBoV group 3 [72]. It is a cost-effective, time-saving method which allows for the detection of multiple targets in one reaction. The main disadvantage is the requirement for the careful optimization of primer concentrations and annealing temperatures to avoid non-specific amplification. Loop-mediated isothermal amplification, or LAMP, was created for the quick, sensitive, and specific detection of PBo-likeV. LAMP amplifies DNA at a constant temperature using a set of four to six primers and a strand-displacing DNA polymerase. It has a stated detection limit of about ten copies per reaction and is about 100-times more sensitive than conventional PCR. Crucially, LAMP does not cross-react with major circulating swine viruses, including classical swine fever virus (CSFV), porcine parvovirus (PPV), porcine circovirus type 2 (PCV-2), and pig reproductive and respiratory syndrome virus (PRRSV) [73]. Primer design can be complex, and the method is less suitable for quantification compared to PCR. Targeting the VP1 gene, TaqMan-based real-time PCR (qPCR) has been utilized to detect and quantify ungulate bocavirus 2. TaqMan-based real-time PCR uses a fluorogenic probe to detect and quantify

DNA during PCR amplification. This approach is quite specific and sensitive [68]. The specificity of the TaqMan PCR assay was evaluated using eight different reactions, which included Pbo-likeV, PRRSV, PCV2, pseudorabies virus (PRV), CSFV, Japanese encephalitis virus (JEV), and porcine torque tenovirus (PTTV). By comparing the signal intensity at various levels, Pbo-likeV was easily distinguished from other pig viruses. The sensitivity of the real-time PCR assay was evaluated by testing ten-fold serial dilutions of the DNA standards (2.00×10^1 to 2.00×10^7 copies) [40]. This method is quite expensive due to the cost of probes and requires specialized equipment. An EvaGreen-based multiplex real-time PCR approach was developed in order to identify and group the three BoVs groups (groups 1, 2, and 3) simultaneously in a single step [74]. This method utilizes EvaGreen dye and melting curve analysis to detect multiple targets in a single reaction. It is cost-effective, has high sensitivity, and allows for the detection of multiple pathogens. The main disadvantage is that this method requires careful optimization to avoid non-specific amplification and dye-related artifacts. Duplex nano PCR has been used to amplify and quantify PBoV and PRV simultaneously [75]. Duplex nano PCR combines nanopore sequencing with PCR to achieve high accuracy in detecting and quantifying DNA. The main advantages are its high accuracy, ability to sequence long reads, and suitability for complex samples. As a main disadvantage, the requirement for specialized equipment can be highlighted. SYBR Green-based duplex real-time PCR has been used for the detection of PBoV genotypes (3, 4, and 5) and porcine epidemic diarrhea virus (PEDV), with detection limits of 10 copies/lL for both viruses [76]. This method uses SYBR Green dye to detect and quantify DNA in real-time PCR, allowing for the simultaneous detection of two targets. It is cost-effective, simple to use, and suitable for high-throughput screening. This method is less specific than probe-based methods and can produce non-specific signals. Researchers in Sweden [16] implemented random amplification and high-throughput sequencing (HTS) to identify the PBo-likeV. These techniques are increasingly popular for discovering new viruses and addressing metagenomic challenges [40,66,77,78]. HTS allows for the sequencing of multiple samples simultaneously, generating large amounts of data quickly, but it is expensive and requires extensive data analysis and specialized equipment. Single-Primer Amplification (SISPA) for PBoV detection employs a single primer for sequence-independent PCR amplification of nucleic acids with unknown sequences. Researchers have successfully applied this method to amplify PBoV [19]. SISPA amplifies DNA using a single primer, allowing for the detection of unknown regions adjacent to known sequences. It is useful for genome walking and detecting unknown sequences but can produce non-specific amplification and requires careful optimization. Nested polymerase chain reaction has been used for the identification of PBoV from throat swab, fecal, and serum samples [69,78]. Nested polymerase chain reaction uses two sets of primers in two successive PCR reactions to increase specificity and sensitivity. It features high specificity, reduces non-specific amplification, and is suitable for detecting low-abundance targets, but it is more time-consuming and requires additional steps compared to conventional PCR. In the primer walking approach, researchers use specific primers in conjunction with a degenerate primer to target known BoV sequences [58]. The primer walking approach is a sequencing method that uses a series of overlapping primers to sequence long DNA fragments step-by-step. It is suitable for sequencing long DNA fragments and filling gaps in genome sequences, but it is time-consuming, labor-intensive, and requires careful primer design. The availability of additional sequences is essential for designing specific primers. Nucleic acid sequence-based amplification (NASBA), an isothermal amplification technique, assesses the polarity of the human BoV genome and analyzes uncharacterized gene transcription [67].

Indirect immunofluorescence tests have been developed to detect unknown isolated PBoV in primary porcinev kidney cell lines. Staining is observed in the cytoplasm and nuclei of infected cell cultures [21]. Additionally, monoclonal antibodies against PBoV 3 and 4 are produced for antigen-detecting ELISA [79]. An indirect enzyme-linked immunosorbent assay (ELISA) based on recombinant nucleoprotein 1 (NP1) investigates PBoV seroprevalence in China. Importantly, there is no cross-reactivity with antiserum against PRRSV, PCV-2, PPV, PEDV, or transmissible gastroenteritis virus (TGEV) [80].

5. Pathogenesis and Public Health Importance of PBoV

PBoV is quite common in the swine population and exhibits a high level of genetic diversity. However, its exact pathogenesis remains undetermined. Clinical evidence suggests that PBoV's pathogenesis could be linked to direct disease manifestations [12]. Notably, PBoV has been detected in various tissues, indicating a wide tissue tropism [60]. Initial detection occurred in the lymph nodes of weaning pigs with post-weaning multisystemic wasting syndrome (PMWS) [16]. Subsequently, PBoV was found in respiratory tract samples and fecal samples from piglets [19]. In cases of gastrointestinal clinical signs, histopathological alterations (microscopic lesions and villous atrophy) were primarily found in the jejunum, ileum, and duodenum [81]. Additionally, the pathogenic role of PBoV has been highlighted by its isolation from piglets suffering from encephalomyelitis [24]. A study also identified a new porcine parvovirus (PPV4), similar to PBoV, in lung lavage from a diseased pig co-infected with PCV-2. Inoculation of colostrum-deprived pigs with lung, lymph nodes, spleen, and heart tissue homogenates resulted in respiratory disease and euthanasia.

The pathophysiology of PBoV is still unknown, despite its great frequency in the swine population and considerable genetic variety [31]. While PBoV has been detected in various tissues, including lymph nodes, spleen, and tonsils, it remains unclear which specific cells in pigs contribute to its replication and dissemination throughout the body [30]. Notably, PBoV has also been found in saliva and serum [22,28,82]. In a study conducted with dogs in Thailand, researchers reported that most viruses from the *Parvoviridae* family replicate in mitotically active cells, such as intestinal crypt epithelial cells. Examination of tissue samples from the small intestine revealed nuclear signals for canine bocavirus 2 primarily in enterocytes located at villus tips and crypts. Transmission electron microscopy showed electron-dense icosahedral viral particles measuring approximately 20 nm in diameter, creating large intranuclear inclusion bodies within apical small intestinal enterocytes [83]. These findings shed light on canine bocavirus infections' pathogenicity and may have implications for other BoVs isolated from different animals. Interestingly, rats with a parvovirus infection [26] generally remain healthy. But, occasionally, pregnant rats experimentally inoculated with parvovirus exhibit growth retardation and fetal loss. Despite these insights, cell lines for BoVs' isolation have not yet been reported.

For PBoV propagation, various cell types have been used, for example human embryonic kidney epithelial cells (HEK293T), monkey kidney cells (MARC-145), porcine kidney cells (PK-15), porcine testicular cells, and porcine alveolar macrophages. However, despite these efforts, no successful propagation has been reported [59].

During a diarrheal disease outbreak in Belgium, researchers attempted to isolate recombinant enterovirus-torovirus and BoV from porcine testicular cells and primary porcine kidney epithelial cells. No virus was detected in passaged cells, despite the fact that after two days a cytopathic effect was seen in pig kidney epithelial cells [25].

Another study conducted in Northern Ireland reported that after four passes in cultured primary swine kidney cells, fecal suspensions and homogenates from the small intestines of 6-week-old piglets showed a cytopathic effect. Interestingly, PCR and RT-PCR did not find porcine enterovirus types 1, 2, or 3, porcine adenovirus, porcine reovirus, PCV1, PCV-2, or PPV. Successful BoV proliferation in a primary cell line was originally reported in this study [21]. However, due to the virus's inability to be cultured in cell lines and its common coexistence with other circulating swine enteric viruses, its precise role in disease progression remains unclear. The association of BoV's with other pathogenic enteric viruses has perpetually confounded its nature. Additionally, the dual involvement of BoVs in respiratory tract infections and gastrointestinal tract infections raises questions about whether they primarily act as respiratory or enteric pathogens, especially in humans, pigs, and other animals. Unfortunately, the lack of an experimental animal model has hindered investigations into the pathogenic nature of PBoV.

The fecal–oral route may serve as the primary transmission pathway for PBoV, similar to other enteric pathogens. Diarrheal feces, vomitus, and other contaminated fomites (such as feeds and transportation vehicles) are important sources of transmission. As a nonenveloped virus, PBoV is resistant to severe environmental conditions and helps transmit other parvoviruses widely [27]. In 2007, human BoVs were detected in 81% of raw sewage samples from several US states [84], highlighting the robustness of BoVs. PBoV can spread across the swine population and induce subclinical infections in pigs once it is introduced into swine herds. In order to determine whether aerosolized PBoV is infectious, further investigation is needed.

Human BoV human parvovirus 4 and parvovirus B19, both belonging to the *Parvoviridae* family, are associated with human diseases [85,86]. Human BoVs have a global presence [87]. The first documented case of human BoV occurred in a child's respiratory tract in Sweden in 2005, four years before BoV was identified in swine [52]. BoVs have been found in both the respiratory tract [53,84,88] and gastrointestinal tract [89], as well as in cancer patients [90]. These viruses have also been detected in healthy individuals. However, their role in human health remains unknown, since they may act as true pathogens, opportunistic pathogens, and commensal flora. The description of PBoV in a three-year-old kid suffering from an acute respiratory tract infection in an Iranian case study from 2018 [14] raises the possibility that PBoV is evolving as a human pathogen. The risk of interspecies transmission of PBoV to human subjects can arise when xenotransplantation is carried out [86]. Therefore, given the growing concern in regard to the pathogenic role of PBoV, it is recommended to screen and analyze the transmission risk of ssDNA viruses to prevent post-transplantation complications.

The possibility of interspecies transmission from rats and pigs to humans requires more research [49]. Significant public health concerns are raised by the vast host range, wide tissue tropism, and growing pathogenic potential of BoVs.

6. Economic Impact and Prevention

PBoV has significant economic implications for the swine industry and food industry. It can lead to reduced growth rates and increased mortality in piglets, which directly impacts productivity. Infected pigs often exhibit symptoms such as diarrhea and respiratory issues, leading to lower weight gain, higher feed conversion ratios, and increased mortality rates, all of which negatively impact farm profitability [35]. For instance, studies have shown that co-infection with PBoV and other pathogens, such as PRRSV, can exacerbate these effects, leading to even greater economic losses [91]. A study on the impact of PBoV and PRRSV co-infection found that pigs with the worst production performance had significantly higher levels of co-infection, resulting in increased mortality and lower weight gain [92]. The study highlighted that infected pigs showed a reduction in daily live weight gain (DLWG) and deteriorating feed conversion ratios (FCRs), which are critical indicators of production efficiency [92]. Managing PBoV involves substantial veterinary expenses. Farmers need to invest in diagnostic tests, treatments, and preventive measures to control the spread of the virus. This includes costs for medications, vaccines, and biosecurity measures [70]. Outbreaks of PBoV can lead to trade restrictions and export bans, especially if the virus is detected in countries that are major exporters of pork. These restrictions can result in substantial financial losses for producers and affect the overall market dynamics, as well as prices of meat [35]. Overall, the economic impact of PBoV is multifaceted, affecting productivity, increasing costs, and disrupting trade. Addressing these challenges requires coordinated efforts in disease management and prevention.

Preventing PBoV outbreaks involves several key strategies. Implementing strict biosecurity protocols is crucial. This includes controlling access to farms, disinfecting equipment and vehicles, and ensuring that workers follow hygiene practices to prevent the introduction and spread of the virus [71]. While there is no specific vaccine for PBoV yet, keeping pigs vaccinated against other common diseases can help maintain overall herd health, reducing the risk of secondary infections that can exacerbate PBoV symptoms [35]. Conducting regular health checks and diagnostic tests can help detect PBoV infection in an early stage of introduction to the farm. Early detection allows for timely intervention, reducing the spread of the virus within the herd [90]. Isolating new or returning animals for a period before introducing them to the main herd can prevent the introduction of PBoV from external sources [71]. Ensuring proper nutrition, reducing stress, and maintaining clean living conditions can strengthen the pigs' immune systems, making them less susceptible to infections [35]. Keeping farm staff informed about PBoV and its prevention is essential. Regular training sessions on biosecurity and disease management can help maintain high standards of farm hygiene and health [91]. By combining these strategies, farmers can significantly reduce the risk of PBoV outbreaks and protect their herds from this economically impactful virus.

International and national legal provisions play a crucial role in regulating the export of pigs to prevent the spread of diseases, including PBoV. The movement of porcine animals within the EU is regulated by Regulation (EU) 2016/429 of the European Parliament and of the Council (Animal Health Law, AHL) on transmissible animal diseases. This regulation aims to prevent the spread of animal diseases by setting strict biosecurity measures and health requirements for the transport of porcine animals [93]. Additionally, Commission Delegated Regulation (EU) 2020/688 lays down specific health guarantees for the movement of porcine animals to ensure they do not pose a significant risk of spreading diseases [93]. The UK has implemented restrictions on the import of fresh porcine meat and meat products from areas within the EU affected by African Swine Fever (ASF). These restrictions are outlined in various health certificates and regulations, which require that porcine meat products from ASF-affected areas undergo specific treatments to mitigate the risk of disease transmission [94]. International trade rules, such as those enforced by the World Trade Organization (WTO), allow for import bans to be applied to specific regions or establishments affected by disease outbreaks. This principle, known as "regionalization," ensures that trade can continue from disease-free regions within exporting countries, thereby preventing the spread of diseases like PBoV [95]. These regulations highlight the importance of stringent health measures and biosecurity protocols in the export of pigs to prevent the spread of diseases, including PBoV. By adhering to these legal provisions, countries can safeguard their livestock industries and protect public health.

7. Conclusions

BoV, an emerging pathogen with a broad host range and wide tissue tropism, has challenged researchers over the past decade. Despite numerous attempts to isolate this virus in cell culture, success has remained elusive. Consequently, an organ co-culture system may prove beneficial for BoV propagation. Such a system could shed light on whether the virus predominantly targets the respiratory or gastrointestinal tract or other tissue systems.

Numerous studies from different countries have connected PBoV to swine coinfecting diseases such as rotavirus A, PEDV, PRRSV, and CSFV. But even ten years after its description, we are still unsure if PBoV is a real pathogen or just an opportunistic one. To address this, single-virus isolation could be employed in animal infection models. Direct proof of the pathophysiology of PBoV is also essential.

Complex evolutionary pathways, interspecies transmission, and recombination sites have been uncovered by phylogenetic analysis of PBoV isolated from mice and minks. To clarify interspecies transmission across several host species, more investigation is required. Given its potential impact, BoV warrants attention as a significant public health problem. An Iranian clinical case [14] serves as a reminder of the importance of understanding emerging pathogens like PBoV. Our contributions to PBoV research will aid in prevention, control, and mitigating its economic consequences.

The economic ramifications of PBoV distribution are complex. Although the full pathogenic potential of PBoV remains unclear, its interaction with other viral infections can intensify disease severity in pigs, leading to higher rates of illness and death. This situation can cause substantial financial losses for the swine industry due to reduced productivity, increased veterinary expenses, and the necessity for stricter biosecurity protocols. Additionally, the detection of PBoV in pig populations can impact international trade, as countries might enforce restrictions on the import and export of pigs and pork products from affected areas. Such measures can limit market access and result in financial setbacks for pig farmers and associated industries.

In summary, the extensive spread of PBoV and its possible role in disease outbreaks highlight the critical need for ongoing research, vigilant monitoring, and the development of effective control strategies to safeguard the global swine industry and minimize economic losses.

Author Contributions: Conceptualization J.P.; writing—original draft preparation, J.P.; writing—review and editing, L.J., T.K., M.K. and M.B.; supervision L.J. All authors have read and agreed to the published version of this manuscript.

Funding: This research received no external funding.

Institutional Review Board Statement: Not applicable.

Informed Consent Statement: Not applicable.

Data Availability Statement: No new data were created or analyzed in this study. Data sharing is not applicable to this article.

Conflicts of Interest: The authors declare no conflicts of interest.

References

1. González, N.; Marquès, M.; Nadal, M.; Domingo, J.L. Meat consumption: Which are the current global risks? A review of recent (2010–2020) evidences. *Food Res. Int.* **2020**, *137*, 109341. [CrossRef]
2. Augère-Granier, M.L. The EU Pig Meat Sector. Available online: https://www.europarl.europa.eu/RegData/etudes/BRIE/2020/652044/EPRS_BRI(2020)652044_EN.pdf (accessed on 4 June 2024).
3. Ito, S.; Bosch, J.; Martínez-Avilés, M.; Sánchez-Vizcaíno, J.M. The Evolution of African Swine Fever in China: A Global Threat? *Front. Vet. Sci.* **2022**, *9*, 828498. [CrossRef]
4. Simmonds, L. Croatian Pig Farming Industry Threatened with Collapse? Available online: https://total-croatia-news.com/news/business/croatian-pig-farming/ (accessed on 2 June 2024).
5. Grgić, I.; Levak, V.; Zrakić, M. Croatian competitiveness in the production of pig meat before joining the EU. In Proceedings of the AAEM 10th International Conference, Skopje, Republic of Macedonia, 12–14 May 2016.
6. OEC. Pig Meat in Croatia. Available online: https://oec.world/en/profile/bilateral-product/pig-meat/reporter/hrv (accessed on 5 July 2024).
7. Drew, T.W. The emergence and evolution of swine viral diseases: To what extent have husbandry systems and global trade contributed to their distribution and diversity? *Rev. Sci. Tech. OIE* **2011**, *30*, 95–106. [CrossRef]
8. Nathues, H.; Alarcon, P.; Rushton, J.; Jolie, R.; Fiebig, K.; Jimenez, M.; Geurts, V. Cost of porcine reproductive and respiratory syndrome virus at individual farm level–An economic disease model. *Prev. Vet. Med.* **2017**, *142*, 16–29. [CrossRef]
9. Baekbo, P.; Kristensen, C.S.; Larsen, L.E. Porcine Circovirus Diseases: A review of PMWS. *Transbound. Emerg. Dis.* **2012**, *59*, 60–67. [CrossRef]
10. Vidigal, P.M.; Mafra, C.L.; Silva, F.M.; Fietto, J.I.; Júnior, A.S.; Almeida, M.R. Tripping over emerging pathogens around the world: A phylogeographical approach for determining the epidemiology of Porcine circovirus-2 (PCV-2), considering global trading. *Virus Res.* **2012**, *163*, 320–327. [CrossRef] [PubMed]
11. Gaudreault, N.N.; Madden, D.W.; Wilson, W.C.; Trujillo, J.D.; Richt, J.A. African Swine Fever Virus: An Emerging DNA Arbovirus. *Front. Vet. Sci.* **2020**, *7*, 215. [CrossRef]
12. Zhou, Y.; Xu, J.; Zhu, S.K.; Meng, Q.F.; Lin, Z.X.; Chen, R.; Qian, A.D. Genetic analysis of three porcine bocaparvoviruses and identification of a natural recombinant breakpoint in NS1. *Arch. Virol.* **2017**, *163*, 707–712. [CrossRef] [PubMed]
13. Zhang, J.; Lu, Y.; Ku, S.L.X.; Liu, X.; Memon, A.M.; He, Q.; Bi, D.; Meng, X. Co-infection with porcine bocavirus and porcine circovirus 2 affects inflammatory cytokine production and tight junctions of IPEC-J2 cells. *Virus Genes* **2018**, *54*, 684–693. [CrossRef] [PubMed]
14. Safamanesh, S.; Azimian, A.; Shakeri, A.; Ghazvini, K.; Amel Jamehdar, S.; Khosrojerdi, M.; Youssefi, M. Detection of porcine bocavirus from a child with acute respiratory tract infection. *Pediatric Infect. Dis. J.* **2018**, *37*, e338–e339. [CrossRef]

15. Keros, T.; Jemeršić, L.; Toplak, I.; Prpić, J. The silent spread of Porcine Bocavirus in Croatian pigs: Should we be concerned? *Acta Vet. Hung.* **2017**, *65*, 565–573. [CrossRef]
16. Blomström, A.L.; Belák, S.; Fossum, C.; McKillen, J.; Allan, G.; Wallgren, P.; Berg, M. Detection of a novel porcine boca-like virus in the background of porcine circovirus type 2 induced postweaning multisystemic wasting syndrome. *Virus Res.* **2009**, *146*, 125–129. [CrossRef]
17. Manteufel, J.; Truyen, U. Animal bocaviruses: A brief review. *Intervirology* **2008**, *51*, 328–334. [CrossRef] [PubMed]
18. Zhai, S.; Yue, C.; Wei, Z.; Long, J.; Ran, D.; Lin, T.; Yuan, S. High prevalence of a novel porcine bocavirus in weanling piglets with respiratory tract symptoms in China. *Arch. Virol.* **2010**, *155*, 1313–1317. [CrossRef]
19. Cheng, W.X.; Li, J.S.; Huang, C.P.; Yao, D.P.; Liu, N.; Cui, S.X.; Jin, Y.; Duan, Z.J. Identification and nearly full-length genome characterization of novel porcine bocaviruses. *PLoS ONE* **2010**, *5*, e13583. [CrossRef] [PubMed]
20. Cadar, D.; Cságola, A.; Lőrincz, M.; Tombácz, K.; Kiss, T.; Spînu, M.; Tuboly, T. Genetic detection and analysis of porcine bocavirus type 1 (PoBoV1) in European wild boar (Sus scrofa). *Virus Genes* **2011**, *43*, 376–379. [CrossRef]
21. McKillen, J.; McNeilly, F.; Duffy, C.; McMenamy, M.; McNair, I.; Hjertner, B.; Millar, A.; McKay, K.; Lagan, P.; Adair, B.; et al. Isolation in cell cultures and initial characterisation of two novel bocavirus species from swine in Northern Ireland. *Vet. Microbiol.* **2011**, *152*, 39–45. [CrossRef]
22. Csagola, A.; Lorincz, M.; Cadar, D.; Tombacz, K.; Biksi, I.; Tuboly, T. Detection, prevalence and analysis of emerging porcine parvovirus infections. *Arch. Virol.* **2012**, *157*, 1003–1010. [CrossRef] [PubMed]
23. Vlasakova, M.; Leskova, V.; Sliz, I.; Jackova, A.; Vilcek, S. The presence of six potentially pathogenic viruses in pigs suffering from post-weaning multisystemic wasting syndrome. *BMC Vet. Res.* **2014**, *10*, 221. [CrossRef] [PubMed]
24. Pfankuche, V.M.; Bodewes, R.; Hahn, K.; Puff, C.; Beineke, A.; Habierski, A.; Osterhaus, A.D.M.E.; Baumgärtner, W. Porcine bocavirus infection associated with encephalomyelitis in a pig, Germany. *Emerg. Infect. Dis.* **2016**, *22*, 1310–1312. [CrossRef]
25. Conceicao-Neto, N.; Theuns, S.; Cui, T.; Zeller, M.; Yinda, C.K.; Christiaens, I.; Heylen, E.; Van Ranst, M.; Carpentier, S.; Nauwynck, H.J.; et al. Identification of an enterovirus recombinant with a torovirus-like gene insertion during a diarrhea outbreak in fattening pigs. *Virus Evol.* **2017**, *3*, vex024. [CrossRef]
26. McMenamy, M.; McKillen, J.; McNair, I.; Duffy, C.; Blomström, A.L.; Charreyre, C.; Welsh, M.; Allan, G. Detection of a porcine boca-like virus in combination with porcine circovirus type 2 genotypes and Torque teno sus virus in pigs from postweaning multisystemic wasting syndrome (PMWS)-affected and nonPMWS-affected farms in archival samples from Great Britain. *Vet. Microbiol.* **2013**, *164*, 293–298. [PubMed]
27. Saekhow, P.; Ikeda, H. Prevalence and genomic characterization of porcine parvoviruses detected in Chiangmai area of Thailand in 2011. *Microbiol. Immunol.* **2014**, *59*, 82–88. [CrossRef] [PubMed]
28. Choi, M.G.; Park, S.J.; Nguyen, V.G.; Chung, H.C.; Kim, A.R.; Park, B.K. Molecular detection and genetic analysis of porcine bocavirus in Korean domestic swine herds. *Arch. Virol.* **2014**, *159*, 1487–1492. [CrossRef]
29. Zhang, W.; Sano, N.; Kataoka, M.; Ami, Y.; Suzaki, Y.; Wakita, T.; Ikeda, H.; Li, T.C. Virus-like particles of porcine bocavirus generated by recombinant baculoviruses can be applied to sero-epidemic studies. *Virus Res.* **2016**, *217*, 85–91. [CrossRef] [PubMed]
30. Jacob, D.M.; Lee, C.Y.; Arshad, S.S.; Selvarajah, G.T.; Bande, F.; Ong, B.L.; Ooi, P.T. First molecular detection of porcine bocavirus in Malaysia. *Trop. Anim. Health Prod.* **2018**, *50*, 733–739. [CrossRef]
31. Cheung, A.K.; Wu, G.; Wang, D.; Bayles, D.O.; Lager, K.M.; Vincent, A.L. Identification and molecular cloning of a novel porcine parvovirus. *Arch. Virol.* **2010**, *155*, 801–806. [CrossRef] [PubMed]
32. Schirtzinger, E.E.; Suddith, A.W.; Hause, B.M.; Hesse, R.A. First identification of porcine parvovirus 6 in North America by viral metagenomic sequencing of serum from pigs infected with porcine reproductive and respiratory syndrome virus. *Virol. J.* **2015**, *12*, 170. [CrossRef]
33. Ndze, V.N.; Cadar, D.; Csagola, A.; Kisfali, P.; Kovacs, E.; Farkas, S.; Ngu, A.F.; Esona, M.D.; Dan, A.; Tuboly, T.; et al. Detection of novel porcine bocaviruses in fecal samples of asymptomatic pigs in Cameroon. *Infect. Genet. Evol.* **2013**, *17*, 277–282. [CrossRef] [PubMed]
34. Amimo, J.O.; Njuguna, J.; Machuka, E.; Okoth, E.; Djikeng, A. First complete genome sequences of porcine bocavirus strains from East Africa. *Genome Announc.* **2017**, *5*, e00093-17. [CrossRef]
35. Aryal, M.; Liu, G. Porcine Bocavirus: A 10-Year History since Its Discovery. *Virol Sin.* **2021**, *36*, 1261–1272. [CrossRef] [PubMed]
36. Mapchart. Available online: https://www.mapchart.net/world.html (accessed on 8 October 2024).
37. International Committee on Taxonomy of Viruses: ICTV. Available online: https://ictv.global/report/chapter/parvoviridae/parvoviridae/bocaparvovirus (accessed on 4 June 2024).
38. Cotmore, S.F.; Agbandje-McKenna, M.; Canuti, M.; Chiorini, J.A.; EisHubinger, A.M.; Hughes, J.; Mietzsch, M.; Modha, S.; Ogliastro, M.; Penzes, J.J.; et al. ICTV virus taxonomy profile: Parvoviridae. *J. Gen. Virol.* **2019**, *100*, 367–368. [CrossRef] [PubMed]
39. Lau, S.K.P.; Woo, P.C.Y.; Tse, H.; Fu, C.T.Y.; Au, W.K.; Chen, X.C.; Tsoi, H.W.; Tsang, T.H.F.; Chan, J.S.Y.; Tsang, D.N.C.; et al. Identification of novel porcine and bovine parvoviruses closely related to human parvovirus 4. *J. Gen. Virol.* **2008**, *89*, 1840–1848. [CrossRef] [PubMed]
40. Li, L.; Shan, T.; Wang, C.; Côté, C.; Kolman, J.; Onions, D.; Gulland, F.M.; Delwart, E. The fecal viral flora of California sea lions. *J. Virol.* **2011**, *85*, 9909–9917. [CrossRef]

41. Kapoor, A.; Mehta, N.; Dubovi, E.J.; Simmonds, P.; Govindasamy, L.; Medina, J.L.; Street, C.; Shields, S.; Lipkin, W.I. Characterization of novel canine bocaviruses and their association with respiratory disease. *J. Gener. Virol.* **2012**, *93*, 341–346. [CrossRef] [PubMed]
42. Li, L.; Pesavento, P.A.; Leutenegger, C.M.; Estrada, M.; Cofey, L.L.; Naccache, S.N.; Samayoa, E.; Chiu, C.; Qiu, J.; Wang, C.; et al. A novel bocavirus in canine liver. *Virol. J.* **2013**, *10*, 54. [CrossRef]
43. Lau, S.K.P.; Woo, P.C.Y.; Yeung, H.C.; Teng, J.L.L.; Wu, Y.; Bai, R.; Fan, R.Y.Y.; Chan, K.H.; Yuen, K.Y. Identifcation and characterization of bocaviruses in cats and dogs reveals a novel feline bocavirus and a novel genetic group of canine bocavirus. *J. Gener. Virol.* **2012**, *93*, 1573–1582. [CrossRef] [PubMed]
44. Ng, T.F.F.; Mesquita, J.R.; Nascimento, M.S.J.; Kondov, N.O.; Wong, W.; Reuter, G.; Knowles, N.J.; Vega, E.; Esona, M.D.; Deng, X.; et al. Feline fecal virome reveals novel and prevalent enteric viruses. *Vet. Microbiol.* **2014**, *171*, 102–111. [CrossRef] [PubMed]
45. Zhang, W.; Li, L.; Deng, X.; Kapusinszky, B.; Pesavento, P.A.; Delwart, E. Faecal virome of cats in an animal shelter. *J. Gener. Virol.* **2014**, *95*, 2553–2564. [CrossRef] [PubMed]
46. Yang, S.; Wang, Y.; Li, W.; Fan, Z.; Jiang, L.; Lin, Y.; Fu, X.; Shen, Q.; Sun, Z.; Wang, X.; et al. A novel bocavirus from domestic mink. *China Virus Genes* **2016**, *52*, 887–890. [CrossRef]
47. Wu, Z.; Ren, X.; Yang, L.; Hu, Y.; Yang, J.; He, G.; Zhang, J.; Dong, J.; Sun, L.; Du, J.; et al. Virome analysis for identifcation of novel mammalian viruses in bat species from Chinese Provinces. *J. Virol.* **2012**, *86*, 10999–11012. [CrossRef] [PubMed]
48. He, B.; Li, Z.; Yang, F.; Zheng, J.; Feng, Y.; Guo, H.; Li, Y.; Wang, Y.; Su, N.; Zhang, F.; et al. Virome profling of bats from myanmar by metagenomic analysis of tissue samples reveals more novel mammalian viruses. *PLoS ONE* **2013**, *8*, e61950.
49. Lau, S.K.P.; Ahmed, S.S.; Yeung, H.C.; Li, K.S.M.; Fan, R.Y.Y.; Cheng, T.Y.C.; Cai, J.P.; Wang, M.; Zheng, B.J.; Wong, S.S.Y.; et al. Identifcation and interspecies transmission of a novel bocaparvovirus among diferent bat species in China. *J. Gener. Virol.* **2016**, *97*, 3345–3358. [CrossRef]
50. Lau, S.K.P.; Yeung, H.C.; Li, K.S.M.; Lam, C.S.F.; Cai, J.P.; Yuen, M.C.; Wang, M.; Zheng, B.J.; Woo, P.C.; Yuen, K.Y. Identifcation and genomic characterization of a novel rat bocavirus from brown rats in China. *Infect. Genet. Evol.* **2017**, *47*, 68–76. [CrossRef] [PubMed]
51. Lanave, G.; Martella, V.; Farkas, S.L.; Marton, S.; Fehér, E.; Bodnar, L.; Lavazza, A.; Decaro, N.; Buonavoglia, C.; Bányai, K. Novel bocaparvoviruses in rabbits. *Vet. J.* **2015**, *206*, 131–135. [CrossRef] [PubMed]
52. Allander, T.; Tammi, M.T.; Eriksson, M.; Bjerkner, A.; Tiveljung-Lindell, A.; Andersson, B. Cloning of a human parvovirus by molecular screening of respiratory tract samples. *Proc. Natl. Acad. Sci. USA* **2005**, *102*, 12891–12896. [CrossRef]
53. Arthur, J.L.; Higgins, G.D.; Davidson, G.P.; Givney, R.C.; Ratclif, R.M. A novel bocavirus associated with acute gastroenteritis in Australian children. *PLoS Pathog.* **2009**, *5*, e1000391. [CrossRef] [PubMed]
54. Ao, Y.; Duan, Z. Novel primate bocaparvovirus species 3 identifed in wild macaca Mulatta in China. *Virol. Sin.* **2020**, *35*, 34–42. [CrossRef]
55. Williams, S.H.; Che, X.; Garcia, J.A.; Klena, J.D.; Lee, B.; Muller, D.; Ulrich, W.; Corrigan, R.M.; Nichol, S.; Jain, K.; et al. Viral diversity of house mice in New York City. *mBio* **2018**, *9*, e01354-17. [CrossRef] [PubMed]
56. Abinanti, F.R.; Warfeld, M.S. Recovery of a hemadsorbing virus (HADEN) from the gastrointestinal tract of calves. *Virology* **1961**, *14*, 288–289. [CrossRef]
57. Mitra, N.; Cernicchiaro, N.; Torres, S.; Li, F.; Hause, B.M. Metagenomic characterization of the virome associated with bovine respiratory disease in feedlot cattle identifed novel viruses and suggests an etiologic role for infuenza D virus. *J. Gener. Virol.* **2016**, *97*, 1771–1784. [CrossRef] [PubMed]
58. Zeng, S.; Wang, D.; Fang, L.; Ma, J.; Song, T.; Zhang, R.; Chen, H.; Xiao, S. Complete coding sequences and phylogenetic analysis of porcine bocavirus. *J. Gener. Virol.* **2011**, *92*, 784–788. [CrossRef] [PubMed]
59. Shan, T.; Lan, D.; Li, L.; Wang, C.; Cui, L.; Zhang, W.; Hua, X.; Zhu, C.; Zhao, W.; Delwart, E. Genomic characterization and high prevalence of bocaviruses in swine. *PLoS ONE* **2011**, *6*, e17292. [CrossRef]
60. Lau, S.K.P.; Woo, P.C.Y.; Yip, C.C.Y.; Li, K.S.M.; Fu, C.T.Y.; Huang, Y.; Chan, K.H.; Yuen, K.Y. Co-existence of multiple strains of two novel porcine bocaviruses in the same pig, a previously undescribed phenomenon in members of the family Parvoviridae, and evidence for inter- and intra-host genetic diversity and recombination. *J. Gen. Virol.* **2011**, *92*, 2047–2059. [CrossRef] [PubMed]
61. Woo, P.C.Y.; Lau, S.K.P.; Tsoi, H.W.; Patteril, N.G.; Yeung, H.C.; Joseph, S.; Wong, E.Y.M.; Muhammed, R.; Chow, F.W.N.; Wernery, U.; et al. Two novel dromedary camel bocaparvoviruses from dromedaries in the Middle East with unique genomic features. *J. Gener. Virol.* **2017**, *98*, 1349–1359. [CrossRef]
62. Kumar, D.; Chaudhary, S.; Lu, N.; Duf, M.; Hefel, M.; McKinney, C.A.; Bedenice, D.; Marthaler, D. Metagenomic next-generation sequencing reveal presence of a novel ungulate bocaparvovirus in alpacas. *Viruses* **2019**, *11*, 701. [CrossRef]
63. Wang, Y.; Zhao, J.; Zheng, M.; Liu, Z.; Yuan, J.; Zhao, J.; Shen, Q.; Fan, Z.; Jiang, L.; Yang, S. Genome sequence of a porcine bocavirus detected in feces of domestic minks in China. *Genome Announc.* **2017**, *5*, e01170. [CrossRef] [PubMed]
64. Lau, S.K.P.; Syed, S.A.; Hoi-Wah, T.; Hazel, C.Y.; Kenneth, S.M.L.; Rachel, Y.Y.F.; Pyrear, S.H.Z.; Candy, C.C.L.; Carol, S.F.L.; Kelvin, K.F.C.; et al. Bats host diverse parvoviruses as possible origin of mammalian dependoparvoviruses and source for bat–swine interspecies transmission. *J. Gen. Virol.* **2017**, *98*, 3046–3059. [CrossRef]
65. Zhang, C.; Song, F.; Xiu, L.; Liu, Y.; Yang, J.; Yao, L.; Peng, J. Identification and characterization of a novel rodent bocavirus from different rodent species in China. *Emerg. Microbes Infect.* **2018**, *7*, 48. [CrossRef]

66. Yang, W.Z.; Yu, J.M.; Li, J.S.; Cheng, W.X.; Huang, C.P.; Duan, Z.J. Genome characterization of a novel porcine bocavirus. *Arch. Virol.* **2012**, *157*, 2125–2132. [CrossRef] [PubMed]
67. Böhmer, A.; Schildgen, V.; Lüsebrink, J.; Ziegler, S.; Tillmann, R.L.; Kleines, M.; Schildgen, O. Novel application for isothermal nucleic acid sequence-based amplification (NASBA). *J. Virol. Methods* **2009**, *158*, 199–201. [CrossRef]
68. Zhou, Y.; Xu, J.; Wang, W.L.; Song, S.W.; Zhu, S.K.; Meng, Q.F.; Yu, F.; Li, C.P.; Liu, N.; Luan, W.M. A TaqMan-based real-time PCR assay for the detection of ungulate bocaparvovirus 2. *J. Virol. Methods* **2018**, *261*, 17–21. [CrossRef]
69. Sun, Y.; Chen, A.Y.; Cheng, F.; Guan, W.; Johnson, F.B.; Qiu, J. Molecular characterization of infectious clones of the minute virus of canines reveals unique features of bocaviruses. *J. Virol.* **2009**, *83*, 3956–3967. [CrossRef] [PubMed]
70. Zheng, L.L.; Cui, J.T.; Qiao, H.; Li, X.S.; Li, X.K.; Chen, H.Y. Detection and genetic characteristics of porcine bocavirus in central China. *Arch. Virol.* **2021**, *166*, 451–460. [CrossRef] [PubMed]
71. Zhou, F.; Sun, H.; Wang, Y. Porcine Bocavirus: Achievements in the Past Five Years. *Viruses* **2014**, *6*, 4946–4960. [CrossRef]
72. Zheng, X.; Liu, G.; Opriessnig, T.; Wang, Z.; Yang, Z.; Jiang, Y. Development and validation of a multiplex conventional PCR assay for simultaneous detection and grouping of porcine bocaviruses. *J. Virol. Methods* **2016**, *236*, 164–169. [CrossRef]
73. Li, B.; Ma, J.J.; Xiao, S.B.; Zhang, X.H.; Wen, L.B.; Mao, L.; Ni, Y.X.; Guo, R.L.; Zhou, J.M.; Lv, L.X.; et al. Development of a loop-mediated isothermal amplification method for rapid detection of porcine boca-like virus. *J. Virol. Methods* **2013**, *179*, 390–395. [CrossRef] [PubMed]
74. Zheng, X.; Liu, G.; Opriessnig, T.; Wang, Z.; Yang, Z.; Jiang, Y. Rapid detection and grouping of porcine bocaviruses by an EvaGreen((R)) based multiplex real-time PCR assay using melting curve analysis. *Mol. Cell. Probes* **2016**, *30*, 195–204. [CrossRef] [PubMed]
75. Luo, Y.; Liang, L.; Zhou, L.; Zhao, K.; Cui, S. Concurrent infections of pseudorabies virus and porcine bocavirus in China detected by duplex nanoPCR. *J. Virol. Methods* **2015**, *219*, 46–50. [CrossRef]
76. Zheng, L.L.; Cui, J.T.; Han, H.Y.; Hou, H.L.; Wang, L.; Liu, F.; Chen, H.Y. Development of a duplex SYBR Green based real-time PCR assay for detection of porcine epidemic diarrhea virus and porcine bocavirus3/4/5. *Mol. Cell Probes* **2020**, *51*, 101544. [CrossRef]
77. Shan, T.; Li, L.; Simmonds, P.; Wang, C.; Moeser, A.; Delwart, E. The fecal virome of pigs on a high-density farm. *J. Virol.* **2011**, *85*, 11697–11708. [CrossRef]
78. Xiong, Y.Q.; You, F.F.; Chen, X.J.; Chen, Y.X.; Wen, Y.Q.; Chen, Q. Detection and phylogenetic analysis of porcine bocaviruses carried by murine rodents and house shrews in China. *Trans. Bound. Emerg. Dis.* **2018**, *66*, 259–267. [CrossRef]
79. McNair, I.; McNeilly, F.; Duffy, C.; McKillen, J.; McMenamy, M.; Welsh, M.; Allan, G. Production, characterisation and applications of monoclonal antibodies to two novel porcine bocaviruses from swine in Northern Ireland. *Arch. Virol.* **2011**, *156*, 2157–2162. [CrossRef] [PubMed]
80. Shi, Q.K.; Zhang, J.L.; Gu, W.Y.; Hou, L.S.; Yuan, G.F.; Chen, S.J.; Fan, J.H.; Zuo, Y.Z. Seroprevalence of porcine bocavirus in pigs in north-central China using a recombinant-NP1-protein-based indirect ELISA. *Arch. Virol.* **2019**, *164*, 2351–2354. [CrossRef] [PubMed]
81. Zhang, Q.; Hu, R.; Tang, X.; Wu, C.; He, Q.; Zhao, Z.; Chen, H.; Wu, B. Occurrence and investigation of enteric viral infections in pigs with diarrhea in China. *Arch. Virol.* **2013**, *158*, 1631–1636. [CrossRef] [PubMed]
82. Meng, Q.; Qiao, M.; Gong, S.; Tian, L.; Li, C.; Qiao, J.; Meng, D.; Wu, Y.; Cai, K.; Zhang, Z.; et al. Molecular detection and genetic diversity of porcine bocavirus in piglets in China. *Acta Virol.* **2018**, *62*, 343–349. [CrossRef]
83. Piewbang, C.; Wardhani, S.W.; Phongroop, K.; Lohavicharn, P.; Sirivisoot, S.; Kasanul, T.; Techangamsuwan, S. Naturally acquired feline bocavirus type 1 and 3 infections in cats with neurologic deficits. *Transbound. Emerg. Dis.* **2022**, *69*, e3076–e3087. [CrossRef]
84. Blinkova, O.; Rosario, K.; Li, L.; Kapoor, A.; Slikas, B.; Bernardin, F.; Breitbart, M.; Delwart, E. Frequent detection of highly diverse variants of cardiovirus, cosavirus, bocavirus, and circovirus in sewage samples collected in the United States. *J. Clin. Microbiol.* **2009**, *47*, 3507–3513. [CrossRef] [PubMed]
85. Martin, E.T.; Taylor, J.; Kuypers, J.; Magaret, A.; Wald, A.; Zerr, D.; Englund, J.A. Detection of bocavirus in saliva of children with and without respiratory illness. *J. Clin. Microbiol.* **2009**, *47*, 4131–4132. [CrossRef]
86. Karuppannan, A.K.; Opriessnig, T. Possible risks posed by single-stranded DNA viruses of pigs associated with xenotransplantation. *Xenotransplantation* **2018**, *25*, e12453. [CrossRef] [PubMed]
87. Rikhotso, M.C.; Kabue, J.P.; Ledwaba, S.E.; Traore, A.N.; Potgieter, N. Prevalence of human bocavirus in africa and other developing countries between 2005 and 2016: A potential emerging viral pathogen for diarrhea. *J. Trop. Med.* **2018**, 7875482. [CrossRef]
88. Vicente, D.; Cilla, G.; Montes, M.; Pérez-Yarza, E.G.; Pérez-Trallero, E. Human bocavirus, a respiratory and enteric virus. *Emerg. Infect. Dis.* **2007**, *13*, 636–637. [CrossRef] [PubMed]
89. Zhao, H.; Zhao, L.; Sun, Y.; Qian, Y.; Liu, L.; Jia, L.; Zhang, Y.; Dong, H. Detection of a bocavirus circular genome in fecal specimens from children with acute diarrhea in Beijing, China. *PLoS ONE* **2012**, *7*, e48980. [CrossRef]
90. Schildgen, V.; Malecki, M.; Tillmann, R.L.; Brockmann, M.; Schildgen, O. The human bocavirus is associated with some lung and colorectal cancers and persists in solid tumors. *PLoS ONE* **2013**, *8*, e68020. [CrossRef]
91. Barber, E.; Leedom Larson, K. Porcine Bocavirus. Swine Health Information Center and Center for Food Security and Pubic Health. 2016. Available online: http://www.cfsph.iastate.edu/pdf/shic-factsheet-porcine-bocavirus (accessed on 8 October 2024).

92. Rodrigues da Costa, M.; García Manzanilla, E.; Diana, A.; van Staaveren, N.; Torres-Pitarch, A.; Boyle, L.A.; Calderón Díaz, J.A. Identifying challenges to manage body weight variation in pig farms implementing all-in-all-out management practices and their possible implications for animal health: A case study. *Porc. Health Manag.* **2021**, *7*, 10. [CrossRef] [PubMed]
93. European Commission—Food Safety. Available online: https://food.ec.europa.eu/animals/live-animal-movements/porcine-animals_en (accessed on 11 December 2024).
94. European Commission—Trade. Available online: https://policy.trade.ec.europa.eu/eu-trade-relationships-country-and-region/exporters-stories/germany-battling-barriers-enabling-exports-pork-south-korea_en (accessed on 11 December 2024).
95. GOV.UK—Imports and EU Policy Team. Available online: http://apha.defra.gov.uk/documents/bip/ovs-notes/2024-24.pdf (accessed on 11 December 2024).

MDPI AG
Grosspeteranlage 5
4052 Basel
Switzerland
Tel.: +41 61 683 77 34

Veterinary Sciences Editorial Office
E-mail: vetsci@mdpi.com
www.mdpi.com/journal/vetsci

www.ingramcontent.com/pod-product-compliance
Lightning Source LLC
Chambersburg PA
CBHW062058220926
51850CB00010B/2813
* 9 7 8 3 7 2 5 8 6 1 0 9 5 *